Principles of
Sedimentology

Principles of Sedimentology

Edited by
Cody Long

Larsen & Keller
www.larsen-keller.com

Principles of Sedimentology
Edited by Cody Long
ISBN: 978-1-63549-254-5 (Hardback)

☰ Larsen & Keller

Published by Larsen and Keller Education,
5 Penn Plaza,
19th Floor,
New York, NY 10001, USA

Cataloging-in-Publication Data

Principles of sedimentology / edited by Cody Long.
    p. cm.
Includes bibliographical references and index.
ISBN 978-1-63549-254-5
1. Sedimentology. 2. Sedimentation and deposition.
3. Geology, Stratigraphic. 4. Sediments (Geology). I. Long, Cody.
QE471 .P74 2017
552.5--dc23

This book contains information obtained from authentic and highly regarded sources. All chapters are published with permission under the Creative Commons Attribution Share Alike License or equivalent. A wide variety of references are listed. Permissions and sources are indicated; for detailed attributions, please refer to the permissions page. Reasonable efforts have been made to publish reliable data and information, but the authors, editors and publisher cannot assume any responsibility for the vailidity of all materials or the consequences of their use.

Trademark Notice: All trademarks used herein are the property of their respective owners. The use of any trademark in this text does not vest in the author or publisher any trademark ownership rights in such trademarks, nor does the use of such trademarks imply any affiliation with or endorsement of this book by such owners.

The publisher's policy is to use permanent paper from mills that operate a sustainable forestry policy. Furthermore, the publisher ensures that the text paper and cover boards used have met acceptable environmental accreditation standards.

Printed and bound in the United States of America.

For more information regarding Larsen and Keller Education and its products, please visit the publisher's website www.larsen-keller.com

Table of Contents

Preface   **VII**

Chapter 1   **Introduction to Sedimentology**   **1**

Chapter 2   **Sediment: An Overview**   **6**
    a. Sediment   6
    b. Grain Size   7
    c. Bedform   18
    d. Aeolian Processes   22
    e. Exner Equation   27

Chapter 3   **Sedimentary Rocks and Structures**   **53**
    a. Sedimentary Rock   53
    b. Sedimentary Structures   110
    c. Cross-bedding   116
    d. Dish Structure   121
    e. Ripple Marks   123
    f. Sole Markings   128
    g. Soft-sediment Deformation Structures   131

Chapter 4   **Processes Involved in Sedimentology**   **135**
    a. Sediment Transport   135
    b. Deposition (Geology)   155
    c. Erosion   160
    d. Diagenesis   169
    e. Weathering   172

Chapter 5   **Sedimentary Basin: An Integrated Study**   **184**
    a. Sedimentary Basin   184
    b. Basin Modelling   208
    c. Sedimentary Basin Analysis   208

Chapter 6   **Stratigraphy: An Integrated Study**   **210**
    a. Stratigraphy   210
    b. Law of Superposition   210
    c. Principle of Original Horizontality   211
    d. Cross-cutting Relationships   212

Chapter 7   **Applications of Sedimentary Rocks**   **219**
    a. Ceramic   219
    b. Dimension Stone   234

c. Petroleum Geology   246
d. Aquifer   250

Chapter 8 **Allied Fields of Sedimentology**   **259**
a. Geology   259
b. Petrology   279

Permissions

Index

Preface

Sedimentology refers to the study of different sediments present on planet Earth. This study includes the examination of clay, silt and sand so as to get information regarding the erosion process, lithification, deposition, weathering and diagenesis, etc. This book is a compilation of chapters that discuss the fundamental concepts in the field of sedimentology. It includes a detailed explanation of the various methods and techniques of this subject. For all those who are interested in this area, the text can prove to be an essential guide. As this field is emerging at a rapid pace, the contents of this book will help the readers understand the modern concepts and applications of the field.

A short introduction to every chapter is written below to provide an overview of the content of the book:

Chapter 1 - Sedimentology is the study of sediments such as silt, sand and clay. Sedimentary rocks have four types which are clastic rocks, carbonates, evaporates and chemical sedimentary rocks. This section will provide an integrated understanding of sedimentology; **Chapter 2** - Sediment is material found on Earth that is broken down by the process of weathering or erosion. Sediments are usually transported by water but they can also be transported by wind. Some of the topics related to sediments are grain size, bedform, Aeolian and Exner equation. The chapter is an overview of the subject matter incorporating all the major aspects of sediment; **Chapter 3** - Sedimentary rocks are rocks that are formed by the deposition of Earth's surface. The sediment is in its initial stage formed by weathering and erosion from where it is transported to the place of deposition by water or wind. Sedimentary rocks can be classified into clastic rocks, phosphorite, iron-rich sedimentary rocks and carbonate rocks. The topics discussed in the section are of great importance to broaden the existing knowledge on sedimentary rocks; **Chapter 4** - The processes involved in sedimentology are sediment transport, deposition, erosion, diagenesis, weathering etc. Deposition is the process in which sediments are added to a landmass whereas erosion is the process in which soil is removed from the surface of the Earth by wind or water flow and is transported to another location. This section serves as a source to understand the process involved in sedimentology; **Chapter 5** - Sedimentary basins are areas filled with sediments. Some of the examples of sedimentary basin are Angola Basin, Los Angeles Basin, Nias Basin and Niger Delta Basin. This chapter helps the readers in developing an in-depth understanding of sedimentary basins; **Chapter 6** - The chapter deals with the major principles of stratigraphy. Some of these principles are law of superposition, principle of original horizontality and cross-cutting relationships. The law of superposition is a principle that forms the bases of geology, archaeology and some other fields also. The principle of original horizontality states that layers of sediments are deposited horizontally because of gravity. The aspects explained in this section are of vital importance and provides a better understanding of sedimentology; **Chapter 7** - The applications of sedimentary rocks are ceramics, dimension stones, petroleum geology and aquifer. Ceramic is an inorganic material that encompasses of metal or nonmetal atoms. These atoms are held in ionic and covalent bonds. This section has been carefully written to provide an easy understanding of the applications of sedimentary rocks; **Chapter 8** - Geology and petrology are considered to be the allied fields of sedimentology. Geology is the study of solid Earth whereas petrology is an important branch of geology that studies the structure of rocks and their origin as well. This chapter will provide a glimpse of related fields of sedimentology briefly.

I extend my sincere thanks to the publisher for considering me worthy of this task. Finally, I thank my family for being a source of support and help.

Editor

Introduction to Sedimentology

Sedimentology is the study of sediments such as silt, sand and clay. Sedimentary rocks have four types which are clastic rocks, carbonates, evaporates and chemical sedimentary rocks. This section will provide an integrated understanding of sedimentology.

Sedimentology encompasses the study of modern sediments such as sand, silt, and clay, and the processes that result in their formation (erosion and weathering), transport, deposition and diagenesis. Sedimentologists apply their understanding of modern processes to interpret geologic history through observations of sedimentary rocks and sedimentary structures.

Sedimentary rocks cover up to 75% of the Earth's surface, record much of the Earth's history, and harbor the fossil record. Sedimentology is closely linked to stratigraphy, the study of the physical and temporal relationships between rock layers or strata.

The premise that the processes affecting the earth today are the same as in the past is the basis for determining how sedimentary features in the rock record were formed. By comparing similar features today to features in the rock record—for example, by comparing modern sand dunes to dunes preserved in ancient aeolian sandstones—geologists reconstruct past environments.

Sedimentary Rock Types

There are four primary types of sedimentary rocks: clastics, carbonates, evaporites, and chemical.

- Clastic rocks are composed of particles derived from the weathering and erosion of precursor rocks and consist primarily of fragmental material. Clastic rocks are classified according to their predominant grain size and their composition. In the past, the term "Clastic Sedimentary Rocks" were used to describe silica-rich clastic sedimentary rocks, however there have been cases of clastic carbonate rocks. The more appropriate term is siliciclastic sedimentary rocks.

  o Organic sedimentary rocks are important deposits formed from the accumulation of biological detritus, and form coal and oil shale deposits, and are typically found within basins of clastic sedimentary rocks

- Carbonates are composed of various carbonate minerals (most often calcium carbonate ($CaCO_3$)) precipitated by a variety of organic and inorganic processes. Typically, the majority of carbonate rocks are composed of reef material.

- Evaporites are formed through the evaporation of water at the Earth's surface and most commonly include halite or gypsum.

- Chemical sedimentary rocks, including some carbonates, are deposited by precipitation of minerals from aqueous solution. These include jaspilite and chert.

Importance of Sedimentary Rocks

Mi Vida uranium mine in redox mudstones near Moab, Utah

Sedimentary rocks provide a multitude of products which modern and ancient society has come to utilise.

- Art: marble, although a metamorphosed limestone, is an example of the use of sedimentary rocks in the pursuit of aesthetics and art

- Architectural uses: stone derived from sedimentary rocks is used for dimension stone and in architecture, notably slate, a meta-shale, for roofing, sandstone for load-bearing buttresses

- Ceramics and industrial materials: clay for pottery and ceramics including bricks; cement and lime derived from limestone.

- Economic geology: sedimentary rocks host large deposits of SEDEX ore deposits of lead-zinc-silver, large deposits of copper, deposits of gold, tungsten, Uranium, and many other precious minerals, gemstones and industrial minerals including heavy mineral sands ore deposits

- Energy: petroleum geology relies on the capacity of sedimentary rocks to generate deposits of petroleum oils. Coal and oil shale are found in sedimentary rocks. A large proportion of the world's uranium energy resources are hosted within sedimentary successions.

- Groundwater: sedimentary rocks contain a large proportion of the Earth's groundwater aquifers. Our understanding of the extent of these aquifers and how much water can be withdrawn from them depends critically on our knowledge of the rocks that hold them (the reservoir).

Basic Principles

Heavy minerals (dark) deposited in a quartz beach sand (Chennai, India).

The aim of sedimentology, studying sediments, is to derive information on the depositional conditions which acted to deposit the rock unit, and the relation of the individual rock units in a basin into a coherent understanding of the evolution of the sedimentary sequences and basins, and thus, the Earth's geological history as a whole.

The scientific basis of this is the principle of uniformitarianism, which states that the sediments within ancient sedimentary rocks were deposited in the same way as sediments which are being deposited at the Earth's surface today.

Sedimentological conditions are recorded within the sediments as they are laid down; the form of the sediments at present reflects the events of the past and all events which affect the sediments, from the source of the sedimentary material to the stresses enacted upon them after diagenesis are available for study.

The principle of superposition is critical to the interpretation of sedimentary sequences, and in older metamorphic terrains or fold and thrust belts where sediments are often intensely folded or deformed, recognising younging indicators or graded bedding is critical to interpretation of the sedimentary section and often the deformation and metamorphic structure of the region.

Folding in sediments is analysed with the principle of original horizontality, which states that sediments are deposited at their angle of repose which, for most types of sediment, is essentially horizontal. Thus, when the younging direction is known, the rocks can be "unfolded" and interpreted according to the contained sedimentary information.

The principle of lateral continuity states that layers of sediment initially extend laterally in all directions unless obstructed by a physical object or topography.

The principle of cross-cutting relationships states that whatever cuts across or intrudes into the layers of strata is younger than the layers of strata.

Methodology of Sedimentology

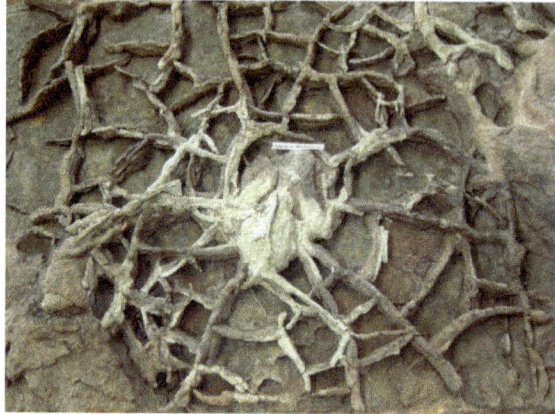

Centripetal desiccation cracks (with a dinosaur footprint in the center) in the Lower Jurassic Moenave Formation at the St. George Dinosaur Discovery Site at Johnson Farm, southwestern Utah.

The methods employed by sedimentologists to gather data and evidence on the nature and depositional conditions of sedimentary rocks include;

- Measuring and describing the outcrop and distribution of the rock unit;
    - Describing the rock formation, a formal process of documenting thickness, lithology, outcrop, distribution, contact relationships to other formations
    - Mapping the distribution of the rock unit, or units
- Descriptions of rock core (drilled and extracted from wells during hydrocarbon exploration)
- Sequence stratigraphy
    - Describes the progression of rock units within a basin
- Describing the lithology of the rock;
    - Petrology and petrography; particularly measurement of texture, grain size, grain shape (sphericity, rounding, etc.), sorting and composition of the sediment
- Analysing the geochemistry of the rock
    - Isotope geochemistry, including use of radiometric dating, to determine the age of the rock, and its affinity to source regions

Recent Developments in Sedimentology

The longstanding understanding of how some mudstones form has been challenged by geologists at Indiana University (Bloomington) and the Massachusetts Institute of Technology. The research, which appears in the December 14th, 2007, edition of *Science*, counters the prevailing view of geologists that mud only settles when water is slow-moving or still, instead showing that "muds will accumulate even when currents move swiftly." The research shows that some mudstones may have formed in fast-moving waters: "Mudstones can be deposited under more energetic conditions than widely assumed, requiring a reappraisal of many geologic records."

Macquaker and Bohacs, in reviewing the research of Schieber et al., state that "these results call for critical reappraisal of all mudstones previously interpreted as having been continuously deposited under still waters. Such rocks are widely used to infer past climates, ocean conditions, and orbital variations."

Considerable recent research into mudstones has been driven by the recent effort to commercially produce hydrocarbons from them, in both the Shale gas and Tight Oil (or Light Tight Oil) plays.

References

- Georges Millot, translated [from the French] by W.R. Farrand, Helene Paquet, Geology Of Clays - Weathering, Sedimentology, Geochemistry Springer Verlag, Berlin (1970), ISBN 0-412-10050-9.

- Gary Nichols, Sedimentology & Stratigraphy, Wiley-Blackwell, Malden, MA (1999), ISBN 0-632-03578-1.

- Donald R. Prothero and Fred Schwab, Sedimentary Geology: An Introduction to Sedimentary Rocks and Stratigraphy, W. H. Freeman (1996), ISBN 0-7167-2726-9.

- Edward J. Tarbuck, Frederick K. Lutgens, Cameron J. Tsujita, Earth, An Introduction to Physical Geology, National Library of Canada Cataloguing in Publication, 2005, ISBN 0-13-121724-0

2

Sediment: An Overview

Sediment is material found on Earth that is broken down by the process of weathering or erosion. Sediments are usually transported by water but they can also be transported by wind. Some of the topics related to sediments are grain size, bedform, Aeolian and Exner equation. The chapter is an overview of the subject matter incorporating all the major aspects of sediment.

Sediment

Sediment is a naturally occurring material that is broken down by processes of weathering and erosion, and is subsequently transported by the action of wind, water, or ice, and/or by the force of gravity acting on the particles. For example, sand and silt can be carried in suspension in river water and on reaching the sea be deposited by sedimentation and if buried this may eventually become sandstone and siltstone, (sedimentary rocks).

River Rhône flowing into Lake Geneva. Sediments make the water appear brownish-grey; they are an indicator of increased water runoff, land degradation, erosion due to intensive industrialized land use, land sealing, and poor soil management.

Sediment billowing out from Italy's shore into the Adriatic Sea

Sediments are most often transported by water (fluvial processes), but also wind (ae-olian processes) and glaciers. Beach sands and river channel deposits are examples of fluvial transport and deposition, though sediment also often settles out of slow-moving or standing water in lakes and oceans. Desert sand dunes and loess are examples of aeolian transport and deposition. Glacial moraine deposits and till are ice-transported sediments.

Classification

Sediment can be classified based on its grain size and/or its composition.

Grain Size

Sediment size is measured on a log base 2 scale, called the "Phi" scale, which classifies particles by size from "colloid" to "boulder".

Sediment in the Gulf of Mexico

Sediment off the Yucatán Peninsula

φ scale	Size range (metric)	Size range (inches)	Aggregate class (Wentworth)	Other names
< -8	> 256 mm	> 10.1 in	Boulder	
-6 to -8	64–256 mm	2.5–10.1 in	Cobble	
-5 to -6	32–64 mm	1.26–2.5 in	Very coarse gravel	Pebble
-4 to -5	16–32 mm	0.63–1.26 in	Coarse gravel	Pebble
-3 to -4	8–16 mm	0.31–0.63 in	Medium gravel	Pebble
-2 to -3	4–8 mm	0.157–0.31 in	Fine gravel	Pebble
-1 to -2	2–4 mm	0.079–0.157 in	Very fine gravel	Granule
0 to -1	1–2 mm	0.039–0.079 in	Very coarse sand	
1 to 0	0.5–1 mm	0.020–0.039 in	Coarse sand	
2 to 1	0.25–0.5 mm	0.010–0.020 in	Medium sand	
3 to 2	125–250 μm	0.0049–0.010 in	Fine sand	
4 to 3	62.5–125 μm	0.0025–0.0049 in	Very fine sand	
8 to 4	3.9–62.5 μm	0.00015–0.0025 in	Silt	Mud
> 8	< 3.9 μm	< 0.00015 in	Clay	Mud
>10	< 1 μm	< 0.000039 in	Colloid	Mud

Composition

Composition of sediment can be measured in terms of:

- parent rock lithology

- mineral composition

- chemical make-up.

This leads to an ambiguity in which clay can be used as both a size-range and a composition.

Sediment Transport

Sediment builds up on human-made breakwaters because they reduce the speed of water flow, so the stream cannot carry as much sediment load.

Glacial transport of boulders. These boulders will be deposited as the glacier retreats.

Sediment is transported based on the strength of the flow that carries it and its own size, volume, density, and shape. Stronger flows will increase the lift and drag on the particle, causing it to rise, while larger or denser particles will be more likely to fall through the flow.

Fluvial Processes: Rivers, Streams, and Overland Flow

Particle Motion

Rivers and streams carry sediment in their flows. This sediment can be in a variety of locations within the flow, depending on the balance between the upwards velocity on the particle (drag and lift forces), and the settling velocity of the particle. These relationships are shown in the following table for the Rouse number, which is a ratio of sediment fall velocity to upwards velocity.

$$\textbf{Rouse} = \frac{\text{Settling velocity}}{\text{Upwards velocity from lift and drag}} = \frac{w_s}{\kappa u_*}$$

where

- w_s is the fall velocity

- κ is the von Kármán constant

- u_* is the shear velocity

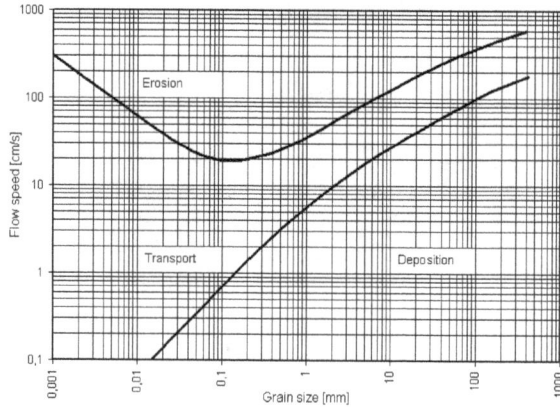

Hjulström curve: The velocities of currents required for erosion, transportation, and deposition (sedimentation) of sediment particles of different sizes.

Mode of Transport	Rouse Number
Bed load	>2.5
Suspended load: 50% Suspended	>1.2, <2.5
Suspended load: 100% Suspended	>0.8, <1.2
Wash load	<0.8

If the upwards velocity approximately equal to the settling velocity, sediment will be transported downstream entirely as suspended load. If the upwards velocity is much less than the settling velocity, but still high enough for the sediment to move, it will move along the bed as bed load by rolling, sliding, and saltating. If the upwards velocity is higher than the settling velocity, the sediment will be transported high in the flow as wash load.

As there are generally a range of different particle sizes in the flow, it is common for material of different sizes to move through all areas of the flow for given stream conditions.

Fluvial Bedforms

Modern asymmetric ripples developed in sand on the floor of the Hunter River, New South Wales, Australia. Flow direction is from right to left.

Sinuous-crested dunes exposed at low tide in the Cornwallis River near Wolfville, Nova Scotia

Ancient channel deposit in the Stellarton Formation (Pennsylvanian), Coalburn Pit, near Thorburn, Nova Scotia.

Sediment motion can create self-organized structures such as ripples, dunes, antidunes on the river or stream bed. These bedforms are often preserved in sedimentary rocks and can be used to estimate the direction and magnitude of the flow that deposited the sediment.

Surface Runoff

Overland flow can erode soil particles and transport them downslope. The erosion associated with overland flow may occur through different methods depending on meteorological and flow conditions.

- If the initial impact of rain droplets dislodges soil, the phenomenon is called rainsplash erosion.

- If overland flow is directly responsible for sediment entrainment but does not form gullies, it is called "sheet erosion".

- If the flow and the substrate permit channelization, gullies may form; this is termed "gully erosion".

Key Fluvial Depositional Environments

The major fluvial (river and stream) environments for deposition of sediments include:

- Deltas (arguably an intermediate environment between fluvial and marine)

- Point bars

- Alluvial fans

- Braided rivers

- Oxbow lakes

- Levees

- Waterfalls

Aeolian Processes: Wind

Wind results in the transportation of fine sediment and the formation of sand dune fields and soils from airborne dust.

Glacial Processes

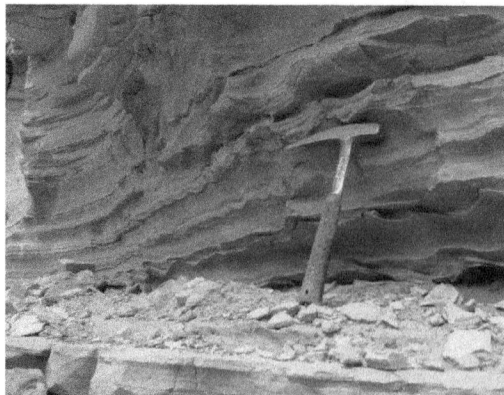

Glacial sediments from Montana

Glaciers carry a wide range of sediment sizes, and deposit it in moraines.

Mass Balance

The overall balance between sediment in transport and sediment being deposited on the bed is given by the Exner equation. This expression states that the rate of increase in bed elevation due to deposition is proportional to the amount of sediment that falls out of the flow. This equation is important in that changes in the power of the flow changes the ability of the flow to carry sediment, and this is reflected in patterns of erosion and deposition observed throughout a stream. This can be localized, and simply due to small obstacles: examples are scour holes behind boulders, where flow accelerates, and deposition on the inside of meander bends. Erosion and deposition can also be regional: erosion can occur due to dam removal and base level fall. Deposition can occur due to dam emplacement that causes the river to pool, and deposit its entire load or due to base level rise.

Shores and Shallow Seas

Seas, oceans and lakes accumulate sediment over time. The sediment could consist of *terrigenous* material, which originates on land, but may be deposited in either terrestrial, marine, or lacustrine (lake) environments; or of sediments (often biological) originating in the body of water. Terrigenous material is often supplied by nearby rivers and streams or reworked marine sediment (e.g. sand). In the mid-ocean, the exoskeletons of dead organisms are primarily responsible for sediment accumulation.

Deposited sediments are the source of sedimentary rocks, which can contain fossils of the inhabitants of the body of water that were, upon death, covered by accumulating sediment. Lake bed sediments that have not solidified into rock can be used to determine past climatic conditions.

Key Marine Depositional Environments

Holocene eolianite and a carbonate beach on Long Island, Bahamas.

The major areas for deposition of sediments in the marine environment include:

- Littoral sands (e.g. beach sands, runoff river sands, coastal bars and spits, largely clastic with little faunal content)

- The continental shelf (silty clays, increasing marine faunal content).

- The shelf margin (low terrigenous supply, mostly calcareous faunal skeletons)

- The shelf slope (much more fine-grained silts and clays)

- Beds of estuaries with the resultant deposits called "bay mud".

One other depositional environment which is a mixture of fluvial and marine is the turbidite system, which is a major source of sediment to the deep sedimentary and abyssal basins as well as the deep oceanic trenches.

Any depression in a marine environment where sediments accumulate over time is known as a sediment trap.

The null point theory explains how sediment deposition undergoes a hydrodynamic sorting process within the marine environment leading to a seaward fining of sediment grain size.

Environmental Issues

Erosion and Agricultural Sediment Delivery to Rivers

One cause of high sediment loads from slash and burn and shifting cultivation of tropical forests. When the ground surface is stripped of vegetation and then seared of all living organisms, the upper soils are vulnerable to both wind and water erosion. In a number of regions of the earth, entire sectors of a country have become erodible. For example, on the Madagascar high central plateau, which constitutes approximately ten percent of that country's land area, most of the land area is devegetated, and gullies have eroded into the underlying soil in furrows typically in excess of 50 meters deep and one kilometer wide. This siltation results in discoloration of rivers to a dark red brown color and leads to fish kills.

Erosion is also an issue in areas of modern farming, where the removal of native vegetation for the cultivation and harvesting of a single type of crop has left the soil unsupported. Many of these regions are near rivers and drainages. Loss of soil due to erosion removes useful farmland, adds to sediment loads, and can help transport anthropogenic fertilizers into the river system, which leads to eutrophication.

Coastal Development and Sedimentation Near Coral Reefs

Watershed development near coral reefs is a primary cause of sediment-related coral stress.The stripping of natural vegetation in the watershed for development exposes soil to increased wind and rainfall, and as a result, could cause exposed sediment to become more susceptible to erosion and delivery to the marine environment during rainfall events. Sediment can negatively affect corals in many ways, such as by physically smothering them, abrading their surfaces, causing corals to expend energy during sediment removal, and causing algal blooms that can ultimately lead to less space on the seafloor where juvenile corals (polyps) can settle.

When sediments are introduced into the coastal regions of the ocean, the proportion of land, marine and organic-derived sediment that characterizes the seafloor near sources of sediment output is altered. In addition, because the source of sediment (i.e. land, ocean, or organically) is often correlated with how coarse or fine sediment grain sizes that characterize an area are on average, grain size distribution of sediment will shift according to relative input of land (typically fine), marine (typically coarse), and organically-derived (variable with age) sediment. These alterations in marine sediment characterize the amount of sediment that is suspended in the water column at any given time and sediment-related coral stress.

Grain Size

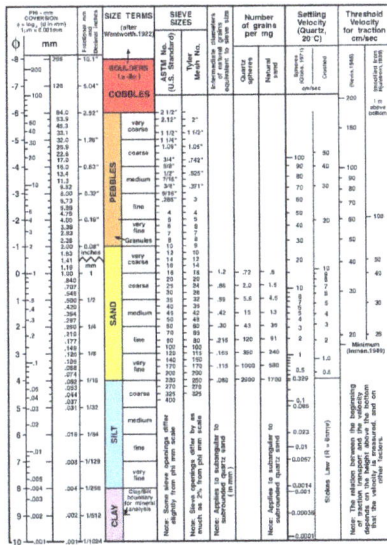

Wentworth grain size chart from United States Geological Survey Open-File Report 2006-1195

Beach cobbles at Nash Point, South Wales.

Particle size, also called grain size, refers to the diameter of individual grains of sediment, or the lithified particles in clastic rocks. The term may also be applied to other granular materials. This is different from the crystallite size, which refers to the size of a single crystal inside a particle or grain. A single grain can be composed of several crystals. Granular material can range from very small colloidal particles, through clay, silt, sand, gravel, and cobbles, to boulders.

Krumbein Phi Scale

Size ranges define limits of classes that are given names in the Wentworth scale (or Udden–Wentworth scale) used in the United States. The Krumbein *phi* (φ) scale, a modification of the Wentworth scale created by W. C. Krumbein in 1937, is a logarithmic scale computed by the equation

$$\phi = -\log_2 D / D_0,$$

where

ϕ is the Krumbein phi scale,

D is the diameter of the particle or grain in millimeters (from petrowiki, krumbein and monks equation) and

D_0 is a reference diameter, equal to 1 mm (to make the equation dimensionally consistent).

This equation can be rearranged to find diameter using φ:

$$D = D_0 \times 2^{-\phi}$$

φ scale	Size range (metric)	Size range (approx. inches)	Aggregate name (Wentworth class)	Other names
<−8	>256 mm	>10.1 in	Boulder	
−6 to −8	64–256 mm	2.5–10.1 in	Cobble	
−5 to −6	32–64 mm	1.26–2.5 in	Very coarse gravel	Pebble
−4 to −5	16–32 mm	0.63–1.26 in	Coarse gravel	Pebble
−3 to −4	8–16 mm	0.31–0.63 in	Medium gravel	Pebble
−2 to −3	4–8 mm	0.157–0.31 in	Fine gravel	Pebble
−1 to −2	2–4 mm	0.079–0.157 in	Very fine gravel	Granule
0 to −1	1–2 mm	0.039–0.079 in	Very coarse sand	
1 to 0	0.5–1 mm	0.020–0.039 in	Coarse sand	
2 to 1	0.25–0.5 mm	0.010–0.020 in	Medium sand	

3 to 2	125–250 µm	0.0049–0.010 in	Fine sand	
4 to 3	62.5–125 µm	0.0025–0.0049 in	Very fine sand	
8 to 4	3.9–62.5 µm	0.00015–0.0025 in	Silt	Mud
10 to 8	0.98–3.9 µm	$3.8{\times}10^{-5}$–0.00015 in	Clay	Mud
20 to 10	0.95–977 nm	$3.8{\times}10^{-8}$–$3.8{\times}10^{-5}$ in	Colloid	Mud

In some schemes, gravel is anything larger than sand (comprising granule, pebble, cobble, and boulder in the table above).

International Scale

ISO 14688-1:2002, establishes the basic principles for the identification and classification of soils on the basis of those material and mass characteristics most commonly used for soils for engineering purposes. ISO 14688-1 is applicable to natural soils *in situ*, similar man-made materials *in situ* and soils redeposited by people.

ISO 14688-1:2002				
Name			**Size range (mm)**	**Size range (approx. in)**
Very coarse soil Boulder Cobble	Large boulder	LBo	>630	>24.8031
	Bo	200–630	7.8740–24.803	
	Co	63–200	2.4803–7.8740	
Coarse soil	Gravel	Coarse gravel	CGr	20–63	0.78740–2.4803
		Medium gravel	MGr	6.3–20	0.24803–0.78740
		Fine gravel	FGr	2.0–6.3	0.078740–0.24803
	Sand	Coarse sand	CSa	0.63–2.0	0.024803–0.078740
		Medium sand	MSa	0.2–0.63	0.0078740–0.024803
		Fine sand	FSa	0.063–0.2	0.0024803–0.0078740
Fine soil	Silt	Coarse silt	CSi	0.02–0.063	0.00078740–0.0024803
		Medium silt	MSi	0.0063–0.02	0.00024803–0.00078740
		Fine silt	FSi	0.002–0.0063	0.000078740–0.00024803
	Clay	Cl	≤0.002	≤0.000078740

Sorting

An accumulation of sediment can also be characterized by the grain size distribution. A sediment deposit can undergo sorting when a particle size range is removed by an agency such as a river or the wind. According to a formula the sorting can be quantified as

φ < 0.35	0.35 < φ < 0.50	0.50 < φ < 0.71	0.71 < φ < 1.00	1.00 < φ < 2.00	2.00 < φ
very well sorted	well sorted	moderately well sorted	moderately sorted	poorly sorted	very poorly sorted

Bedform

Current ripples preserved in sandstone of the Moenkopi Formation, Capitol Reef National Park, Utah, United States.

A bedform is a feature that develops at the interface of fluid and a moveable bed, the result of bed material being moved by fluid flow. Examples include ripples and dunes on the bed of a river. Bedforms are often preserved in the rock record as a result of being present in a depositional setting. Bedforms are often characteristic to the flow parameters, and may be used to infer flow depth and velocity, and therefore the Froude number.

Bedforms Initiation

Bedforms are omnipresent in many environments (e.g., fluvial, eolian, glaciofluvial, deltaic and deep sea), although there is still some debate on how they develop. The initiation of bedforms have two - not mutually exclusive - models: defect initiation and instantaneous initiation.

Defect Initiation

The defect theory proposes that the turbulent sweeps that are generated in turbulent flows entrain sediment that upon deposition generates defects in a non-cohesive material. These deposits then propagate downstream via a flow separation process, thus developing bedform fields. The origin of the defects is thought to be linked to packets of hairpin vortex structures. These coherent turbulent structures give rise to entrainment corridors on the mobile bed, forming grain lineations that interact with the low-speed streaks generating an agglomeration of grains. Once a critical height of grains is reached, flow separation occurs over the new structure. Sediment will be eroded close from the reattachment point and deposited downstream creating a new defect. This new defect will thus induce formation of another defect and the process will continue, propagating downstream while the accumulations of grains quickly evolve into small bedforms.

Instantaneous Initiation

In general, the defect propagation theory plays a bigger role at low sediment transport rates since for high rates defects maybe washed away and bedforms generally initiated across the entire bed spontaneously. Venditti et al. (2005) report that instantaneous initiation begins with the formation of a cross-hatch pattern, which leads to chevron-shaped forms that migrate independently of the pattern structure. This chevron-like structure reorganizes to form the future crest lines of the bedforms. Venditti et al. (2006), based on the earlier model by Liu (1957), proposed that instantaneous initiation is a manifestation of an interfacial hydrodynamic instability of Kelvin-Helmholtz type between a highly active pseudofluid sediment layer and the fluid above it. In addition, Venditti et al. (2005) imply that there is no linkage between the instantaneous initiation and coherent turbulent flow structures, since spatially- and temporally-random events should lock in place to generate the cross-hatch pattern. Moreover, there is no clear explanation of the effect of turbulence in the formation of bedforms since bedforms may also occur under laminar flows . It is important to note, that laminar-generated bedform studies used the temporally-averaged flow conditions to determine the degree of turbulence, indicating Reynolds number in the laminar regime. However, instantaneous process, such as burst and sweeps, which are infrequent at low Reynolds number but still present, can be the driving mechanisms to generate the bedforms. The generation of bedforms in laminar flows is still a topic of debate within the scientific community, since if true, it suggests that there should be other processes for defect development other than the one suggested by Best (1992). This alternative model for bedform development at low sediment transport rates should explain the generation of defects and bedforms for cases where the flow is not turbulent.

Bedform Phase Diagrams

Phase or stability diagrams are defined as graphs that show the regimes of existence of one or more stable bed states. The stability of the bed can be defined when the bedform is in equilibrium and does not change in time for the same flow condition. This invariance over time must not be confused with a static morphology or frozen equilibrium; on the contrary, the bed moves and adjusts in a dynamic equilibrium with the flow and sediment transport for that particular condition. These phase diagrams are used for two main purposes: i) for prediction of bed states in a known flow and sediment transport condition, and, ii) as a tool for the reconstruction of paleoenvironments from a known bed state or sedimentary structure. Despite the great utility of such diagrams, they are very difficult to construct, making them either incomplete or very hard to interpret. This complexity lies in the number of variables needed to quantify the system.

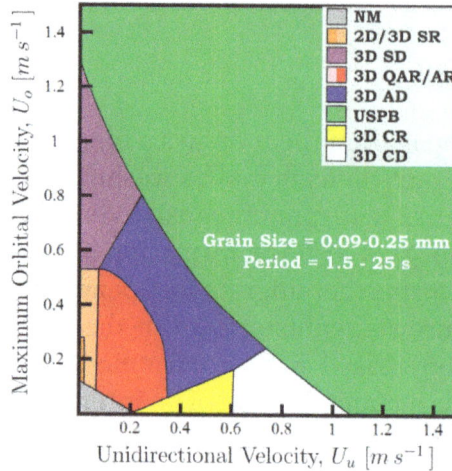

Dimensional phase diagram for combined flows. Relationships of combined-flow bed-phases stability fields in a plot of Oscillatory vs Unidirectional velocity.

Bedforms Vs. Flow

Typical unidirectional bedforms represent a specific flow velocity, assuming typical sediments (sands and silts) and water depths, and a chart such as below can be used for interpreting depositional environments, with increasing water velocity going down the chart.

Flow Regime	Bedform	Preservation Potential	Identification Tips
Lower	Lower plane bed	High	Flat laminae, almost lack of current
	Ripple marks	High	Small, cm-scale undulations
	Sand waves	Medium to low	Rare, longer wavelength than ripples
	Dunes/Megaripples	High	Large, meter-scale ripples
Upper	Upper plane bed	High	Flat laminae, +/- aligned grains (parting lineations)
	Antidunes	Low	Water in phase with bedform, low angle, subtle laminae
	Pool and chute	Very low	Mostly erosional features

This chart is for general use, because changes in grain size and flow depth can change the bedform present and skip bedforms in certain scenarios. Bidirectional environments (e.g. tidal flats) produce similar bedforms, but the reworking the sediments and opposite directions of flow complicates the structures.

This bed form sequence can also be illustrated diagrammatically:

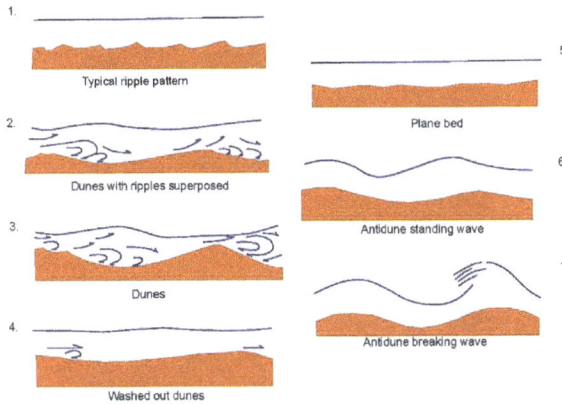

Bedforms formed in sand in channels under unidirectional flow. Numbers correspond broadly to increasing flow regime, i.e., increasing water flow velocity. Blue arrows show schematically flow lines in the water above the bed. Flow is always from left to right.

Types of Bedforms

Lower Plane Bed

"Lower plane bed" refers to the flat configuration the bed of a river that is produced in via low rates of sediment transport.

Upper Plane Bed

Parting lineation, from lower left to upper right; Kayenta Formation, Canyonlands National Park.

"Upper plane bed" features are flat and characterized by a unidirectional flow with high rates of sediment transport as both bed load and suspended load. Upper plane bed conditions can produce parting current lineations, which are typically subtle streaks on the bed surface due to the high energy flow.

Aeolian Processes

Wind erosion of soil at the foot of Chimborazo, Ecuador.

Rock carved by drifting sand below Fortification Rock in Arizona (Photo by Timothy H. O'Sullivan, USGS, 1871)

Aeolian processes, also spelled eolian or æolian, pertain to wind activity in the study of geology and weather and specifically to the wind's ability to shape the surface of the Earth (or other planets). Winds may erode, transport, and deposit materials and are effective agents in regions with sparse vegetation, a lack of soil moisture and a large supply of unconsolidated sediments. Although water is a much more powerful eroding force than wind, aeolian processes are important in arid environments such as deserts.

The term is derived from the name of the Greek god Aeolus, the keeper of the winds.

Wind Erosion

A rock sculpted by wind erosion in the Altiplano region of Bolivia

Sand blowing off a crest in the Kelso Dunes of the Mojave Desert, California.

Wind-carved alcove in the Navajo Sandstone near Moab, Utah

Wind erodes the Earth's surface by deflation (the removal of loose, fine-grained particles by the turbulent action of the wind) and by abrasion (the wearing down of surfaces by the grinding action and sandblasting of windborne particles).

Regions which experience intense and sustained erosion are called deflation zones. Most aeolian deflation zones are composed of desert pavement, a sheet-like surface of rock fragments that remains after wind and water have removed the fine particles. Almost half of Earth's desert surfaces are stony deflation zones. The rock mantle in desert pavements protects the underlying material from deflation.

A dark, shiny stain, called desert varnish or rock varnish, is often found on the surfaces of some desert rocks that have been exposed at the surface for a long period of time. Manganese, iron oxides, hydroxides, and clay minerals form most varnishes and provide the shine.

Deflation basins, called blowouts, are hollows formed by the removal of particles by wind. Blowouts are generally small, but may be up to several kilometers in diameter.

Wind-driven grains abrade landforms. In parts of Antarctica wind-blown snowflakes that are technically sediments have also caused abrasion of exposed rocks. Grinding by particles carried in the wind creates grooves or small depressions. Ventifacts are rocks which have been cut, and sometimes polished, by the abrasive action of wind.

Sculpted landforms, called yardangs, are up to tens of meters high and kilometers long and are forms that have been streamlined by desert winds. The famous Great Sphinx of Giza in Egypt may be a modified yardang.

Transport

Dust storm approaching Spearman, Texas April 14, 1935.

Dust storm in Amarillo, Texas. FSA photo by Arthur Rothstein (1936)

A massive sand storm cloud is about to envelop a military camp as it rolls over Al Asad, Iraq, just before nightfall on April 27, 2005.

Particles are transported by winds through suspension, saltation (skipping or bouncing) and creeping (rolling or sliding) along the ground.

Small particles may be held in the atmosphere in suspension. Upward currents of air support the weight of suspended particles and hold them indefinitely in the surrounding air. Typical winds near Earth's surface suspend particles less than 0.2 millimeters in diameter and scatter them aloft as dust or haze.

Saltation is downwind movement of particles in a series of jumps or skips. Saltation normally lifts sand-size particles no more than one centimeter above the ground and proceeds at one-half to one-third the speed of the wind. A saltating grain may hit other grains that jump up to continue the saltation. The grain may also hit larger grains that are too heavy to hop, but that slowly creep forward as they are pushed by saltating grains. Surface creep accounts for as much as 25 percent of grain movement in a desert.

Aeolian turbidity currents are better known as dust storms. Air over deserts is cooled significantly when rain passes through it. This cooler and denser air sinks toward the desert surface. When it reaches the ground, the air is deflected forward and sweeps up surface debris in its turbulence as a dust storm.

Crops, people, villages, and possibly even climates are affected by dust storms. Some dust storms are intercontinental, a few may circle the globe, and occasionally they may engulf entire planets. When the Mariner 9 spacecraft entered its orbit around Mars in 1971, a dust storm lasting one month covered the entire planet, thus delaying the task of photo-mapping the planet's surface.

Most of the dust carried by dust storms is in the form of silt-size particles. Deposits of this windblown silt are known as loess. The thickest known deposit of loess, 335 meters, is on the Loess Plateau in China. This very same Asian dust is blown for thousands of miles, forming deep beds in places as far away as Hawaii. In Europe and in the Americas, accumulations of loess are generally from 20 to 30 meters thick.

Aeolian transport from deserts plays an important role in ecosystems globally, e.g. by transport of minerals from the Sahara to Amazonia. Saharan dust is also responsible for forming red clay soils in southern Europe. Aeolian processes are affected by human activity, such as the use of 4x4 vehicles.

Small whirlwinds, called dust devils, are common in arid lands and are thought to be related to very intense local heating of the air that results in instabilities of the air mass. Dust devils may be as much as one kilometer high.

Deposition

Cross-bedding of sandstone near Mount Carmel road, Zion National Park, indicating wind action and sand dune formation prior to formation of rock (NPS photo by George A. Grant, 1929)

Mesquite Flat Dunes in Death Valley looking toward the Cottonwood Mountains from the north west arm of Star Dune (2003)

Holocene eolianite deposit on Long Island, The Bahamas. This unit is formed of wind-blown carbonate grains. (2007)

Wind-deposited materials hold clues to past as well as to present wind directions and intensities. These features help us understand the present climate and the forces that molded it. Wind-deposited sand bodies occur as sand sheets, ripples, and dunes.

Sand sheets are flat, gently undulating sandy plots of sand surfaced by grains that may be too large for saltation. They form approximately 40 percent of aeolian depositional surfaces. The Selima Sand Sheet in the eastern Sahara Desert, which occupies 60,000 square kilometers in southern Egypt and northern Sudan, is one of the Earth's largest sand sheets. The Selima is absolutely flat in a few places; in others, active dunes move over its surface.

Wind blowing on a sand surface ripples the surface into crests and troughs whose long axes are perpendicular to the wind direction. The average length of jumps during saltation corresponds to the wavelength, or distance between adjacent crests, of the ripples. In ripples, the coarsest materials collect at the crests causing inverse grading. This distinguishes small ripples from dunes, where the coarsest materials are generally in the troughs. This is also a distinguishing feature between water laid ripples and aeolian ripples.

Wind-blown sand moves up the gentle upwind side of the dune by saltation or creep. Sand accumulates at the brink, the top of the slipface. When the buildup of sand at the brink exceeds the angle of repose, a small avalanche of grains slides down the slipface. Grain by grain, the dune moves downwind.

Accumulations of sediment blown by the wind into a mound or ridge, dunes have gentle upwind slopes on the windward side. The downwind portion of the dune, the lee slope, is commonly a steep avalanche slope referred to as a slipface. Dunes may have more than one slipface. The minimum height of a slipface is about 30 centimeters.

Some of the most significant experimental measurements on aeolian sand movement were performed by Ralph Alger Bagnold, a British engineer who worked in Egypt prior to World War II. Bagnold investigated the physics of particles moving through the atmosphere and deposited by wind. He recognized two basic dune types, the crescentic dune, which he called "barchan," and the linear dune, which he called longitudinal or "seif" (Arabic for "sword").

A 2011 study published in *Catena* examined the effect of vegetation on aeolian dust accumulation in the semiarid steppe of northern China. Using a series of trays with different vegetation coverage and a control model with none, the authors found that an increase in vegetation coverage improves the efficiency of dust accumulation and adds more nutrients to the environment, particularly organic carbon. Two critical point were revealed by their data: 1. the efficiency of trapping dust increases slowly above 15% coverage, and decreases rapidly below 15% coverage. 2. at around 55%-75% coverage, dust accumulation reaches a maximum capacity.

A three year quantitative study on the effects of vegetation removal on wind erosion found that the removal of grasses in an aeolian environment increased the rate of soil deposition. In the same study, a relationship was shown between decreasing plant density with decreasing soil nutrients. Similarly, horizontal soil flux across the test site was shown to increase with increasing vegetation removal.

A 1998 study published in Earth Surfaces Processes and Landforms investigated the relationship between vegetative cover on sand surfaces with the rate of sand transport. It was found that sand flux decreased exponentially with vegetation cover. This was done by measuring plots of land with varying degrees of vegetation against rates of sand transport. The authors contend that this relationship can be utilized to manipulate rates of sediment flux by introducing vegetation in an area or to quantify human impact by recognizing vegetation loss's affect on sandy landscapes.

Exner Equation

The Exner equation is a statement of conservation of mass that applies to sediment in a fluvial system such as a river. It was developed by the Austrian meteorologist and sedimentologist Felix Maria Exner, from whom it derives its name.

The Equation

The Exner equation describes conservation of mass between sediment in the bed of a channel and sediment that is being transported. It states that bed elevation increases (the bed aggrades) proportionally to the amount of sediment that drops out of transport, and conversely decreases (the bed degrades) proportionally to the amount of sediment that becomes entrained by the flow.

Basic Equation

The equation states that the change in bed elevation, η, over time, t, is equal to one over the grain packing density, ε_o, times the negative divergence of sediment flux, q_s.

$$\frac{\partial \eta}{\partial t} = -\frac{1}{\varepsilon_o} \nabla \cdot \mathbf{q_s}$$

Note that ε_o can also be expressed as $(1 - \lambda_p)$, where λ_p equals the bed porosity.

Good values of ε_o for natural systems range from 0.45 to 0.75. A typical good value for spherical grains is 0.64, as given by random close packing. An upper bound for close-packed spherical grains is 0.74048; this degree of packing is extremely improbable in natural systems, making random close packing the more realistic upper bound on grain packing density.

Often, for reasons of computational convenience and/or lack of data, the Exner equation is used in its one-dimensional form. This is generally done with respect to the down-stream direction x, as one is typically interested in the down-stream distribution of erosion and deposition though a river reach.

$$\frac{\partial \eta}{\partial t} = -\frac{1}{\varepsilon_o} \frac{\partial \mathbf{q_s}}{\partial x}$$

Including External Changes in Elevation

An additional form of the Exner equation adds a subsidence term, σ, to the mass-balance. This allows the absolute elevation of the bed η to be tracked over time in a situation in which it is being changed by outside influences, such as tectonic or compression-related subsidence (isostatic compression or rebound). In the convention of the following equation, σ is positive with an increase in elevation over time and is negative with a decrease in elevation over time.

$$\frac{\partial \eta}{\partial t} = -\frac{1}{\varepsilon_o} \nabla \cdot \mathbf{q_s} + \sigma$$

Fluvial Environments for Deposition of Sediment

River Delta

Sacramento (California) Delta at flood stage, early 2009

A river delta is a landform that forms from deposition of sediment carried by a river as the flow leaves its mouth and enters slower-moving or standing water. This occurs where a river enters an ocean, sea, estuary, lake, reservoir, or (more rarely) another river that cannot transport away the supplied sediment.

Despite a popular legend, this use of the word *delta* was not coined by Herodotus.

Formation

River deltas form when a river carrying sediment reaches either (1) a body of water, such as a lake, ocean, or reservoir, (2) another river that cannot remove the sediment quickly enough to stop delta formation, or (3) an inland region where the water spreads out and deposits sediments. The tidal currents also cannot be too strong, as sediment would wash out into the water body faster than the river deposits it. Of course, the river must carry enough sediment to layer into deltas over time. The river's velocity decreases rapidly, causing it to deposit the majority, if not all, of its load. This alluvium builds up to form the river delta. When the flow enters the standing water, it is no longer confined to its channel and expands in width. This flow expansion results in a decrease in the flow velocity, which diminishes the ability of the flow to transport sediment. As a result, sediment drops out of the flow and deposits. Over time, this single channel builds a deltaic lobe (such as the bird's-foot of the Mississippi or Ural river deltas), pushing its mouth into the standing water. As the deltaic lobe advances, the gradient of the river channel becomes lower because the river channel is longer but has the same change in elevation.

As the slope of the river channel decreases, it becomes unstable for two reasons. First, gravity makes the water flow in the most direct course down slope. If the river breaches its natural levees (i.e., during a flood), it spills out onto a new course with a shorter route to the ocean, thereby obtaining a more stable steeper slope. Second, as its slope gets lower, the amount of shear stress on the bed decreases, which results in deposition of sediment within the channel and a rise in the channel bed relative to the floodplain. This makes it easier for the river to breach its levees and cut a new channel that enters the body of standing water at a steeper slope. Often when the channel does this, some of its flow remains in the abandoned channel. When these channel-switching events occur, a mature delta develops a distributary network.

Another way these distributary networks form is from deposition of mouth bars (mid-channel sand and/or gravel bars at the mouth of a river). When this mid-channel bar is deposited at the mouth of a river, the flow is routed around it. This results in additional deposition on the upstream end of the mouth-bar, which splits the river into two distributary channels. A good example of the result of this process is the Wax Lake Delta.

In both of these cases, depositional processes force redistribution of deposition from areas of high deposition to areas of low deposition. This results in the smoothing of the planform (or map-view) shape of the delta as the channels move across its surface and deposit sediment. Because the sediment is laid down in this fashion, the shape of these deltas approximates a fan. The more often the flow changes course, the shape develops as closer to an ideal fan, because more rapid changes in channel position results in more uniform deposition of sediment on the delta front. The Mississippi and Ural River deltas, with their bird's-feet, are examples of rivers that do not avulse often enough to form a symmetrical fan shape. Alluvial fan deltas, as seen by their name, avulse frequently and more closely approximate an ideal fan shape.

Types of Deltas

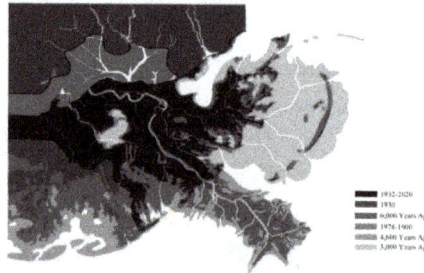

Lower Mississippi River land loss over time

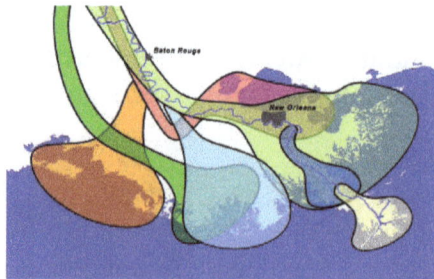

Delta lobe switching in the Mississippi Delta, ■ 4600 yrs BP, ■ 3500 yrs BP, ■ 2800 yrs BP, ■ 1000 yrs BP, ■ 300 yrs BP, ■ 500 yrs BP, ■ current

Deltas are typically classified according to the main control on deposition, which is usually either a river, waves, or tides. These controls have a large effect on the shape of the resulting delta.

Wave-dominated Deltas

In wave dominated deltas, wave-driven sediment transport controls the shape of the delta, and much of the sediment emanating from the river mouth is deflected along the coast line.

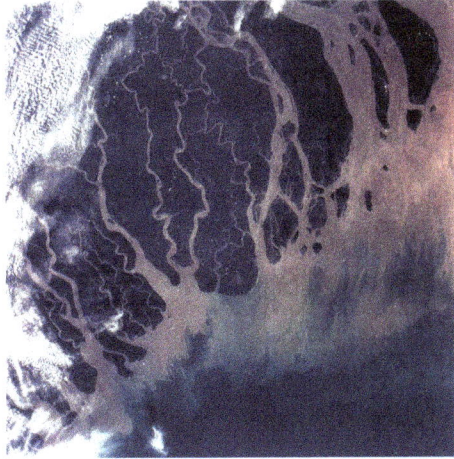

The Ganges Delta in India and Bangladesh is the largest delta in the world and it is also one of the most fertile regions in the world.

Tide-dominated Deltas

Erosion is also an important control in tide dominated deltas, such as the Ganges Delta, which may be mainly submarine, with prominent sand bars and ridges. This tends to produce a "dendritic" structure. Tidal deltas behave differently from river- and wave-dominated deltas, which tend to have a few main distributaries. Once a wave- or river- distributary silts up, it is abandoned, and a new channel forms elsewhere. In a tidal delta, new distributaries are formed during times when there's a lot of water around – such as floods or storm surges. These distributaries slowly silt up at a pretty constant rate until they fizzle out.

Gilbert Deltas

A Gilbert delta (named after Grove Karl Gilbert) is a specific type of delta formed from coarse sediments, as opposed to gently-sloping muddy deltas such as that of the Mississippi. For example, a mountain river depositing sediment into a freshwater lake would form this kind of delta. While some authors describe both lacustrine and marine locations of Gilbert deltas, others note that their formation is more characteristic of the freshwater lakes, where it is easier for the river water to mix with the lakewater faster (as opposed to the case of a river falling into the sea or a salt lake, where less dense fresh water brought by the river stays on top longer).

G.K. Gilbert himself first described this type of delta on Lake Bonneville in 1885. Elsewhere, similar structures occur, for example, at the mouths of several creeks that flow into Okanagan Lake in British Columbia and forming prominent peninsulas at Naramata.

(49°35′30″N 119°35′30″W49.59167°N 119.59167°W), Summerland

(49°34′23″N 119°37′45″W49.57306°N 119.62917°W), or Peachland

(49°47′00″N 119°42′45″W49.78333°N 119.71250°W)

Estuaries

Other rivers, particularly those on coasts with significant tidal range, do not form a delta but enter into the sea in the form of an estuary. Notable examples include the Saint Lawrence River and the Tagus estuary.

Inland Deltas

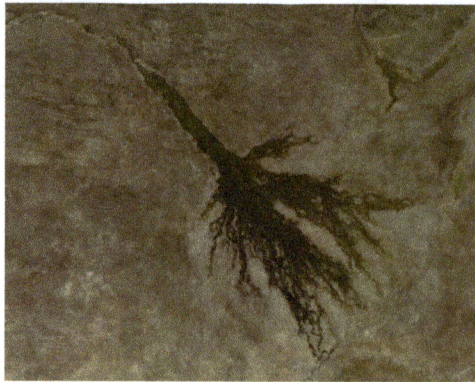

Okavango Delta

In rare cases the river delta is located inside a large valley and is called an inverted river delta. Sometimes a river divides into multiple branches in an inland area, only to rejoin and continue to the sea. Such an area is called an *inland delta*, and often occurs on former lake beds. The Inner Niger Delta and Peace–Athabasca Delta are notable examples. The Amazon has also an inland delta before the island of Marajó.

In some cases, a river flowing into a flat arid area splits into channels that evaporate as it progresses into the desert. Okavango Delta in Botswana is one well-known example.

Sedimentary Structure

The formation of a delta is complicated, multiple, and cross-cutting over time, but in a simple delta three main types of bedding may be distinguished: the bottomset beds, foreset/frontset beds, and topset beds. This three part structure may be seen in small scale by crossbedding.

- The bottomset beds are created from the lightest suspended particles that settle farthest away from the active delta front, as the river flow diminishes into the standing body of water and loses energy. This suspended load is deposited by sediment gravity flow, creating a turbidite. These beds are laid down in horizontal layers and consist of the finest grain sizes.

- The foreset beds in turn are deposited in inclined layers over the bottomset beds as the active lobe advances. Foreset beds form the greater part of the bulk of a delta, (and also occur on the lee side of sand dunes). The sediment particles within foreset beds consist of larger and more variable sizes, and constitute the bed load that the river moves downstream by rolling and bouncing along the channel bottom. When the bed load reaches the edge of the delta front, it rolls over the edge, and is deposited in steeply dipping layers over the top of the existing bottomset beds. Under water, the slope of the outermost edge of the delta is created at the angle of repose of these sediments. As the foresets accumulate and advance, subaqueous landslides occur and readjust overall slope stability. The foreset slope, thus created and maintained, extends the delta lobe outward. In cross section, foresets typically lie in angled, parallel bands, and indicate stages and seasonal variations during the creation of the delta.

- The topset beds of an advancing delta are deposited in turn over the previously laid foresets, truncating or covering them. Topsets are nearly horizontal layers of smaller-sized sediment deposited on the top of the delta and form an extension of the landward alluvial plain. As the river channels meander laterally across the top of the delta, the river is lengthened and its gradient is reduced, causing the suspended load to settle out in nearly horizontal beds over the delta's top. Topset beds are subdivided into two regions: the upper delta plain and the lower delta plain. The upper delta plain is unaffected by the tide, while the boundary with the lower delta plain is defined by the upper limit of tidal influence.

Examples of Deltas

The Ganges/Brahmaputra combination delta which spans most of Bangladesh and empties into the Bay of Bengal, is the world's largest delta. Other rivers with notable deltas include the Nile River, the Fly River, the Godavari, the Krishna, the Kaveri, the Niger River, the Tigris-Euphrates, the Rhine, the Po, the Rhône, the Danube, the Ebro, the Volga, the Lena, the Indus, the Ayeyarwady (Irrawaddy), the Mekong, the Huanghe, the Yangtze, the Sacramento–San Joaquin, the Mississippi, the Orinoco, and the Paraná.

The Ebre River delta at the Mediterranean Sea

The delta of the St. Clair River between the Canadian province of Ontario and the U.S. state of Michigan is the largest delta emptying into a body of fresh water.

Ecological Threats to Deltas

Human activities, including of water and the creation of dams for hydroelectric power or to create reservoirs can radically alter delta ecosystems. Dams block sedimentation, which can cause the delta to erode away. The use of water upstream can greatly increase salinity levels as less fresh water flows to meet the salty ocean water. While nearly all deltas have been impacted to some degree by humans, the Nile Delta and Colorado River Delta are some of the most extreme examples of the ecological devastation caused to deltas by damming and diversion of water.

Deltas in the Economy

Ancient deltas are a benefit to the economy due to their well sorted sand and gravel. Sand and gravel is often quarried from these old deltas and used in concrete for highways, buildings, sidewalks, and even landscaping. More than 1 billion tons of sand and gravel are produced in the United States alone. Not all sand and gravel quarries are former deltas, but for ones that are, a lot of the sorting is already done by the power of water.

As lowlands often adjacent to urban areas, deltas often comprise extensive industrial and commercial areas as well as agricultural land. These uses are often in conflict. The Fraser Delta in British Columbia, Canada, includes the Vancouver Airport and the Roberts Bank Superport and the Annacis Island industrial zone, and a mix of commercial, residential and agricultural land. Space is so limited in the Lower Mainland region, and in British Columbia in general, which is very mountainous, that the Agricultural Land Reserve was created to preserve agricultural land for food production.

Deltas on Mars

Researchers have found a number of examples of deltas that formed in Martian lakes. Finding deltas is a major sign that Mars once had large amounts of water. Deltas have been found over a wide geographical range. Below are pictures of a few.

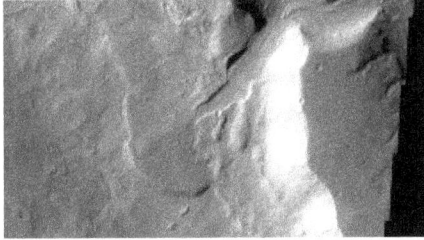

Delta in Ismenius Lacus quadrangle, as seen by THEMIS.

Delta in Lunae Palus quadrangle, as seen by THEMIS.

Delta in Margaritifer Sinus quadrangle as seen by THEMIS.

Point Bar

A point bar is a depositional feature made of alluvium that accumulates on the inside bend of streams and rivers below the slip-off slope. Point bars are found in abundance in mature or meandering streams. They are crescent-shaped and located on the inside of a stream bend, being very similar to, though often smaller than, towheads, or river islands.

Point bars are composed of sediment that is well sorted and typically reflects the overall capacity of the stream. They also have a very gentle *slope* and an elevation very close to water level. Since they are low-lying, they are often overtaken by floods and can accumulate driftwood and other debris during times of high water levels. Due to their near flat topography and the fact that the water speed is slow in the shallows of the point bar they are popular rest stops for boaters and rafters. However, camping on a point bar can be dangerous as a flash flood that raises the stream level by as little as a few inches (centimetres) can overwhelm a campsite in moments.

A point bar is an area of deposition whereas a cut bank is an area of erosion.

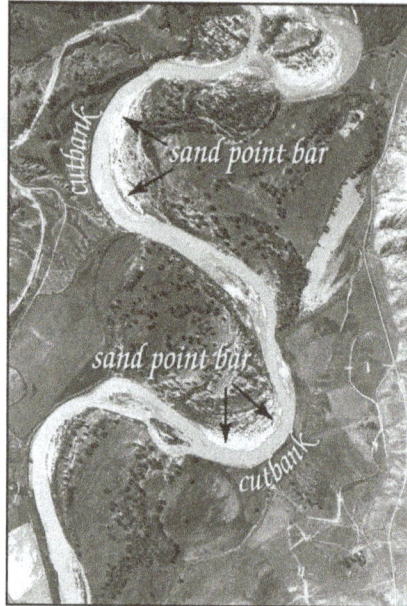

Cut bank erosion and point bar deposition as seen on the Powder River in Montana.

Point bars are formed as the secondary flow of the stream sweeps and rolls sand, gravel and small stones laterally across the floor of the stream and up the shallow sloping floor of the point bar.

Formation

Point bar at a river meander: the Cirque de la Madeleine in the Gorges de l'Ardèche, France.

Any fluid, including water in a stream, can only flow around a bend in vortex flow.

In vortex flow the speed of the fluid is fastest where the radius is smallest, and slowest where the radius is greatest. (Tropical cyclones, tornadoes, and the spinning motion of water as it escapes down a drain are all visible examples of vortex flow.) In the case of water flowing around a bend in a stream the secondary flow in the boundary layer along the floor of the stream does not flow parallel to the banks of the stream but flows partly across the floor of the stream toward the inside of the stream (where the radius of curvature is smallest). This movement of the boundary layer is capable of sweeping and rolling loose particles including sand, gravel, small stones and other submerged objects along the floor of the stream toward the point bar.

This can be demonstrated at home. Partly fill a circular bowl or cup with water and sprinkle a little sand, rice or sugar into the water. Set the water in circular motion with a hand or spoon. The secondary flow will quickly sweep the solid particles into a neat pile in the center of the bowl or cup. The primary flow (the vortex) might be expected to sweep the solid particles to the perimeter of the bowl or cup, but instead the secondary flow along the floor of the bowl or cup sweeps the particles toward the center.

Where a stream is following a straight course the slower boundary layer along the floor of the stream is also following the same straight course. It sweeps and rolls sand, gravel and polished stones downstream, along the floor of the stream. However, as the stream enters a bend and vortex flow commences as the primary flow, a secondary flow also commences and flows partly across the floor of the stream toward the convex bank (the bank with the smaller radius). Sand, gravel and polished stones that have travelled with the stream for a great distance where the stream was following a straight course may finally come to rest in the point bar of the first stream bend.

Due to the circular path of a stream around a bend the surface of the water is slightly higher near the concave bank (the bank with the larger radius) than near the convex bank. This slight slope on the water surface of the stream causes a slightly greater water pressure on the floor of the stream near the concave bank than near the convex bank. This pressure gradient drives the slower boundary layer across the floor of the stream toward the convex bank. The pressure gradient is capable of driving the boundary layer up the shallow sloping floor of the point bar, causing sand, gravel and polished stones to be swept and rolled up-hill.

The concave bank is often a cut bank and an area of erosion. The eroded material is swept and rolled across the floor of the stream by the secondary flow and may be deposited on the point bar only a small distance downstream from its original location in the concave bank.

The point bar typically has a gently sloping floor with shallow water. The shallow water is mostly the accumulated boundary layer and does not have a fast speed.

However, in the deepest parts of the stream where the stream is flowing freely, vortex flow prevails and the stream is flowing fastest where the radius of the bend is smallest, and slowest where the radius is greatest. The shallows around the point bar can become treacherous when the stream is rising. As the water depth increases over the shallows of the point bar, the vortex flow can extend closer toward the convex bank and the water speed at any point can increase dramatically in response to only a small increase in water depth.

Fallacy Regarding Formation of Point Bars

An enduring fallacy exists regarding the formation of point bars and oxbow lakes. The fallacy suggests they are formed by the deposition of a stream's suspended load as the velocity and energy of the stream *decreases* in the bend. This fallacy relies on the erroneous notion that the speed of the water is slowest on the inside of the bend (where the radius is smallest) and fastest on the outside of the bend (where the radius is greatest).

If it were true that, around a bend in a stream, the difference between the speed of the water near one bank and the other was sufficient to cause deposition of suspended solids on one bank but not the other, the deposition would occur near the concave bank rather than the convex bank because vortex flow is slowest where the radius of curvature is greatest. Any fluid, including water in a stream, can only flow around a bend in vortex flow.

All point bars typically have a gently sloping floor with shallow water. The shallow depth of the water, and the fact that it is an accumulation of the boundary layer, prevent the water from reaching fast speed over the point bar. It is probably this observation which led early geographers to believe the slowest part of the stream is where the radius is smallest.

In a slow-flowing stream or river, the difference in speed between one bank and the other is not sufficient to provide a credible explanation as to why all the deposition occurs on one bank, and none on the other. Similarly, the fallacy has no explanation as to why all the deposition occurs at a stream bend, and little or none occurs where the stream is following a straight course. (The speed of water in a stream does not slow just because the stream enters a bend.)

In a mature, meandering stream or river the water speed is slow, turbulence is low, and the water is not capable of holding coarse sand and gravel in suspension. In contrast, point bars comprise coarse sand, gravel, polished stones and other submerged objects. These materials have not been carried in suspension and then dropped on the point bar – they have been swept and rolled into place by the secondary flow that exists across the floor of every stream in the vicinity of a stream bend.

Alluvial Fan

A vast (60 km long) alluvial fan blossoms across the desolate landscape between the Kunlun and Altun mountain ranges that form the southern border of the Taklamakan Desert in Xinjiang. The left side is the active part of the fan, and appears blue from water flowing in the many small streams

An alluvial fan is a fan- or cone-shaped deposit of sediment crossed and built up by streams. If a fan is built up by debris flows it is properly called a debris cone or colluvial fan. These flows come from a single point source at the apex of the fan, and over time move to occupy many positions on the fan surface. Fans are typically found where a canyon draining from mountainous terrain emerges out onto a flatter plain, and especially along fault-bounded mountain fronts.

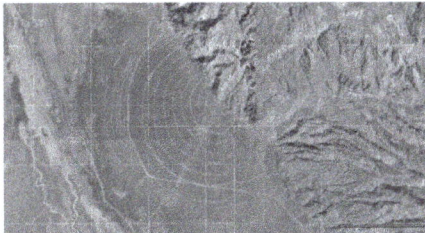

Alluvial fan in Death Valley

Alluvial fan in the French Pyrenees

Alluvial fan above Lake Louise, Alberta, Canada.

A convergence of neighboring alluvial fans into a single apron of deposits against a slope is called a bajada, or compound alluvial fan.

Formation

As a stream's gradient decreases, it drops coarse-grained material. This reduces the capacity of the channel and forces it to change direction and gradually build up a slightly mounded or shallow conical fan shape. The deposits are usually poorly sorted. This fan shape can also be explained with a thermodynamic justification: the system of sediment introduced at the apex of the fan will tend to a state which minimizes the sum of the transport energy involved in moving the sediment and the gravitational potential of material in the fan. There will be iso-transport energy lines forming concentric arcs about the discharge point at the apex of the fan. Thus the material will tend to be deposited equally about these lines, forming the characteristic fan shape.

In Arid Climates

Alluvial fans are often found in desert areas subject to periodic flash floods from nearby thunderstorms in local hills. The typical watercourse in an arid climate has a large, funnel-shaped basin at the top, leading to a narrow defile, which opens out into an alluvial fan at the bottom. Multiple braided streams are usually present and active during water flows.

Phreatophytes are plants that are often concentrated at the base of alluvial fans. They have long tap roots 30 to 50 feet (9.1 to 15.2 m) to reach water that has seeped through the fan and hit an impermeable layer, sometimes collecting in springs and seeps. These stands of shrubs cling to the soil at their bases and often form islands of habitat for many animals as the wind blows the sand around the bushes away.

In Humid Climates

Alluvial fans also develop in wetter climates. In Nepal the Koshi River has built a mega-fan covering some 15,000 km^2 (5,800 sq mi) below its exit from Himalayan foothills onto the nearly level plains where the river traverses into India before joining the Ganges. Along the upper Koshi tributaries, tectonic forces elevate the Himalayas several millimeters annually. Uplift is approximately in equilibrium with erosion, so the river annually carries some 100 million cubic meters (3.5 billion cu ft) of sediment as it exits the mountains. Deposition of this magnitude over millions of years is more than sufficient to account for the megafan.

All along the interface between the Indo-Gangetic Plain and the Himalaya in India, Pakistan, Nepal and Bhutan the outermost, lowest Siwalik foothills are built of poorly consolidated sedimentary rocks that have eroded into a wide, continuous alluvial apron called Bhabar in Hindi and Nepali. Despite overpopulation on the plains, this bhabar zone is highly malarial and has remained largely uninhabited.

In North America, streams flowing into California's Central Valley have deposited smaller but still extensive alluvial fans. Such as that of the Kings River flowing out of the Sierra Nevada creates a low divide, turning the south end of the San Joaquin Valley into an endorheic basin without a connection to the ocean.

Flood Hazards

Alluvial fans are subject to flooding and can be even more dangerous than the upstream canyons that feed them. Their slightly convex perpendicular surfaces cause water to spread widely until there is no zone of refuge. If the gradient is steep, active transport of materials down the fan creates a moving substrate that is inhospitable to travel on foot or wheels. But as the gradient diminishes downslope, water comes down from above faster than it can flow away downstream, and may pond to hazardous depths.

In the case of the Koshi River, the huge sediment load and megafan's slightly convex transverse surface conspire against engineering efforts to contain peak flows inside manmade embankments. In August 2008 high monsoon flows breached the embankment, diverting most of the river into an unprotected ancient channel and across surrounding lands with high population density. Over a million people were rendered homeless, about a thousand lost their lives and thousands of hectares of crops were destroyed. The Koshi is known as the Sorrow of Bihar for contributing disproportionately to India's death tolls in flooding, which exceed those of all countries except Bangladesh.

In the Solar System

Alluvial fans are also found on Mars descending from some crater rims over their flatter floors. Observations of fans in Gale crater made by satellites from orbit have now been confirmed by the discovery of fluvial sediments by the Curiosity rover.

Alluvial fans have been observed by the Cassini-Huygens mission on Titan using the Cassini orbiter's synthetic aperture radar (SAR) instrument. These fans are more common in the drier mid-latitudes at the end of methane/ethane rivers where it is thought that frequent wetting and drying occur due to precipitation, much like arid fans on Earth. Radar imaging suggests that fan material is most likely composed of round grains of water ice or solid organic compounds about two centimetres in diameter.

Levee

A levee, is an elongated naturally occurring ridge or artificially constructed fill or wall, which regulates water levels. It is usually earthen and often parallel to the course of a river in its floodplain or along low-lying coastlines.

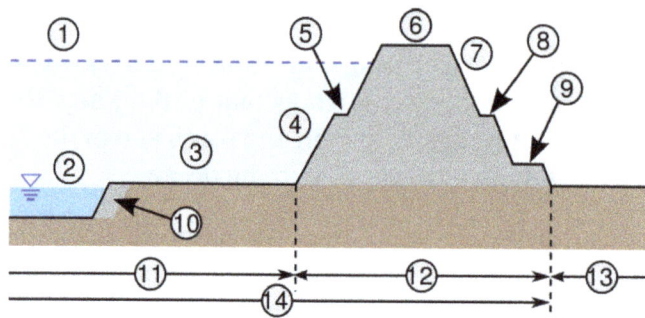

1. Design High Water Level (HWL)

2. Low water channel

3. Flood channel

4. Riverside Slope

5. Riverside Banquette

6. Levee Crown

7. Landside Slope

8. Landside Banquette

9. Berm

10. Low water revetment

11. Riverside land

12. Levee

13. Protected lowland

14. River zone

The side of a levee in Sacramento, California

Etymology

Levee

The word *levee*, from the French word *levée* (from the feminine past participle of the French verb *lever*, "to raise"), is used in American English (notably in the Midwest and Deep South). It originated in New Orleans a few years after the city's founding in 1718 and was later adopted by English speakers. The name derives from the trait of the levee's ridges being *raised* higher than both the channel and the surrounding floodplains.

Dike

The modern word *dike* or *dyke* most likely derives from the Dutch word "*dijk*", with the construction of dikes in Frisia (now part of the Netherlands and Germany) well attested as early as the 11th century. The 126 kilometres (78 mi) long Westfriese Omringdijk was completed by 1250, and was formed by connecting existing older dikes. The Roman chronicler Tacitus even mentions that the rebellious Batavi pierced dikes to flood their land and to protect their retreat (AD 70). The word *dijk* originally indicated both the trench and the bank. It is closely related to the English verb *to dig*.

In Anglo-Saxon, the word *dic* already existed and was pronounced as *dick* in northern England and as *ditch* in the south. Similar to Dutch, the English origins of the word lie in digging a trench and forming the upcast soil into a bank alongside it. This practice has meant that the name may be given to either the excavation or the bank. Thus Offa's Dyke is a combined structure and Car Dyke is a trench though it once had raised banks as well. In the midlands and north of England, and in the United States, a dike is what a ditch is in the south, a property boundary marker or small drainage channel. Where it carries a stream, it may be called a running dike as in *Rippingale Running Dike*, which leads water from the catchwater drain, Car Dyke, to the South Forty Foot Drain in Lincolnshire (TF1427). The Weir Dike is a soak dike in Bourne North Fen, near Twenty and alongside the River Glen, Lincolnshire. In the Norfolk and Suffolk Broads, a dyke may be a drainage ditch or a narrow artificial channel off a river or broad for access or mooring, some longer dykes being named, e.g. Candle Dyke.

In parts of Britain, particularly Scotland, a dyke may be a field wall, generally made with dry stone.

Usage

A reinforced embankment

The main purpose of artificial levees is to prevent flooding of the adjoining countryside and to slow natural course changes in a waterway to provide reliable shipping lanes for maritime commerce over time; they also confine the flow of the river, resulting in higher and faster water flow. Levees can be mainly found along the sea, where dunes are not strong enough, along rivers for protection against high-floods, along lakes or along polders. Furthermore, levees have been built for the purpose of empoldering, or as a boundary for an inundation area. The latter can be a controlled inundation by the military or a measure to prevent inundation of a larger area surrounded by levees. Levees have also been built as field boundaries and as military defences. More on this type of levee can be found in the article on dry-stone walls.

Levees can be permanent earthworks or emergency constructions (often of sandbags) built hastily in a flood emergency. When such an emergency bank is added on top of an existing levee it is known as a *cradge*.

Some of the earliest levees were constructed by the Indus Valley Civilization (in Pakistan and North India from circa 2600 BC) on which the agrarian life of the Harappan peoples depended. Levees were also constructed over 3,000 years ago in ancient Egypt, where a system of levees was built along the left bank of the River Nile for more than 600 miles (970 km), stretching from modern Aswan to the Nile Delta on the shores of the Mediterranean. The Mesopotamian civilizations and ancient China also built large levee systems. Because a levee is only as strong as its weakest point, the height and standards of construction have to be consistent along its length. Some authorities have argued that this requires a strong governing authority to guide the work, and may have been a catalyst for the development of systems of governance in early civilizations. However, others point to evidence of large scale water-control earthen works such as canals and/or levees dating from before King Scorpion in Predynastic Egypt, during which governance was far less centralized.

Levees are usually built by piling earth on a cleared, level surface. Broad at the base, they taper to a level top, where temporary embankments or sandbags can be placed. Because flood discharge intensity increases in levees on both river banks, and because silt deposits raise the level of riverbeds, planning and auxiliary measures are vital. Sections are often set back from the river to form a wider channel, and flood valley basins are divided by multiple levees to prevent a single breach from flooding a large area. A levee made from stones laid in horizontal rows with a bed of thin turf between each of them is known as a *spetchel*.

Artificial levees require substantial engineering. Their surface must be protected from erosion, so they are planted with vegetation such as Bermuda grass in order to bind the earth together. On the land side of high levees, a low terrace of earth known as a *banquette* is usually added as another anti-erosion measure. On the river side, erosion from strong waves or currents presents an even greater threat to the integrity of the levee. The effects of erosion are countered by planting suitable vegetation or installing stones, boulders, weighted matting or concrete revetments. Separate ditches or drainage tiles are constructed to ensure that the foundation does not become waterlogged.

River Flood Prevention

Broken levee on the Sacramento River

Prominent levee systems have been built along the Mississippi River and Sacramento River in the United States, and the Po, Rhine, Meuse River, Rhone, Loire, Vistula, the delta formed by the Rhine, Maas/Meuse and Scheldt in the Netherlands and the Danube in Europe. During the Chinese Warring States period, the Dujiangyan irrigation system was built by the Qin as a water conservation and flood control project. The system's infrastructure is located on the Minjiang, which is the longest tributary of the Chang Jiang, in Sichuan, China.

A levee keeps high water on the Mississippi River from flooding Gretna, Louisiana, in March 2005.

The Mississippi levee system represents one of the largest such systems found anywhere in the world. It comprises over 3,500 miles (5,600 km) of levees extending some 1,000 kilometres (620 mi) along the Mississippi, stretching from Cape Girardeau, Missouri, to the Mississippi Delta. They were begun by French settlers in Louisiana in the 18th century to protect the city of New Orleans. The first Louisiana levees were about 3 feet (0.91 m) high and covered a distance of about 50 miles (80 km) along the riverside. The U.S. Army Corps of Engineers, in conjunction with the Mississippi River Commission, extended the levee system beginning in 1882 to cover the riverbanks from Cairo, Illinois to the mouth of the Mississippi delta in Louisiana. By the mid-1980s, they had reached their present extent and averaged 24 feet (7.3 m) in height; some Mississippi levees are as high as 50 feet (15 m). The Mississippi levees also include some of the longest continuous individual levees in the world. One such levee extends southwards from Pine Bluff, Arkansas, for a distance of some 380 miles (610 km).

Soil Reinforcement and Levee Protection – The United States Army Corps of Engineers (USACE) recommends and supports Cellular Confinement technology (geocells) as a best management practice. Particular attention is given to the matter of surface erosion, overtopping prevention and protection of levee crest and downstream slope. Reinforcement with geocells provides tensile force to the soil to better resist instability.

Effects of Levees upon the Elevation of the River Bed

Artificial levees can lead to an elevation of the natural river bed over time; whether this happens or not and how fast, depends on different factors, one of them being the amount and type of the bed load of a river. Alluvial rivers with intense accumulations of sediment tend to this behavior. Examples of rivers where artificial levees led to an elevation of the river bed, even up to a point where the river bed is higher than the adjacent ground surface behind the levees, are found for the Yellow River in China and the Mississippi in the USA.

Coastal Flood Prevention

Levees are very common on the marshlands bordering the Bay of Fundy in New Bruns-wick and Nova Scotia Canada. The Acadians who settled the area can be credited with the original construction of many of the levees in the area, created for the purpose of farming the fertile tidal marshlands. These levees are referred to as dykes. They are constructed with hinged sluice gates that open on the falling tide to drain freshwater from the agricultural marshlands, and close on the rising tide to prevent seawater from entering behind the dyke. These sluice gates are called "aboiteaux". In the Lower Main-land around the city of Vancouver, British Columbia, there are levees (known locally as dikes, and also referred to as "the sea wall") to protect low-lying land in the Fraser River delta, particularly the city of Richmond on Lulu Island. There are also dikes to protect other locations which have flooded in the past, such as the Pitt Polder, land adjacent to the Pitt River and other tributary rivers.

Coastal flood prevention levees are also common along the inland coastline behind the Wadden Sea, an area devastated by many historic floods. Thus the peoples and govern-ments have erected increasingly large and complex flood protection levee systems to stop the sea even during storm floods. The biggest of these are of course the huge levees in the Netherlands, which have gone beyond just defending against floods, as they have aggressively taken back land that is below mean sea level.

Spur Dykes or Groynes

These typically man-made hydraulic structures are situated to protect against erosion. They are typically placed in alluvial rivers perpendicular, or at an angle, to the bank of the channel or the revetment, and are used widely along coastlines. There are two common types of spur dyke, permeable and impermeable, depending on the materials used to construct them.

Natural Levees

Natural levees commonly form around lowland rivers and creeks without human inter-vention. They are elongate ridges of mud and/or silt that form on the river floodplains immediately adjacent to the cut banks. Like artificial levees, they act to reduce the like-lihood of floodplain inundation.

Deposition of levees is a natural consequence of the flooding of meandering rivers which carry high proportions of suspended sediment in the form of fine sands, silts, and muds. Because the carrying capacity of a river depends in part on its depth, the sed-iment in the water which is over the flooded banks of the channel is no longer capable of keeping the same amount of fine sediments in suspension as the main thalweg. The extra fine sediments thus settle out quickly on the parts of the floodplain nearest to the channel. Over a significant number of floods, this will eventually result in the building up of ridges in these positions, and reducing the likelihood of further floods and episodes of levee building.

If aggradation continues to occur in the main channel, this will make levee overtopping more likely again, and the levees can continue to build up. In some cases this can result in the channel bed eventually rising above the surrounding floodplains, penned in only by the levees around it; an example is the Yellow River in China near the sea, where oceangoing ships appear to sail high above the plain on the elevated river.

Levees are common in any river with a high suspended sediment fraction, and thus are intimately associated with meandering channels, which also are more likely to occur where a river carries large fractions of suspended sediment. For similar reasons, they are also common in tidal creeks, where tides bring in large amounts of coastal silts and muds. High spring tides will cause flooding, and result in the building up of levees.

Levee Failures and Breaches

Both natural and man-made levees can fail in a number of ways. Factors that cause levee failure include overtopping, erosion, structural failures, and levee saturation. The most frequent (and dangerous) is a *levee breach*. Here, a part of the levee actually breaks or is eroded away, leaving a large opening for water to flood land otherwise protected by the levee. A breach can be a sudden or gradual failure, caused either by surface erosion or by subsurface weakness in the levee. A breach can leave a fan-shaped deposit of sediment radiating away from the breach, described as a crevasse splay. In natural levees, once a breach has occurred, the gap in the levee will remain until it is again filled in by levee building processes. This increases the chances of future breaches occurring in the same location. Breaches can be the location of meander cutoffs if the river flow direction is permanently diverted through the gap.

Sometimes levees are said to fail when water *overtops* the crest of the levee. This will cause flooding on the floodplains, but because it does not damage the levee, it has fewer consequences for future flooding.

Among various failure mechanisms that cause levee breaches, soil erosion is found to be one of the most important factors . Predicting soil erosion and scour generation when overtopping happens is important in order to design stable levee and floodwalls. There have been numerous studies to investigate the erodibility of soils. Briaud et al. (2008) used Erosion Function Apparatus (EFA) test to measure the erodibility of the soils and afterwards by using Chen 3D software, numerical simulations were performed on the levee to find out the velocity vectors in the overtopping water and the generated scour when the overtopping water impinges the levee. By analyzing the results from EFA test, an erosion chart to categorize erodibility of the soils was developed. Hughes and Nadal in 2009 studied the effect of combination of wave overtopping and storm surge overflow on the erosion and scour generation in levees. The study included hydraulic parameters and flow characteristics such as flow thickness, wave intervals, surge level above levee crown in analyzing scour development. According to the laboratory tests, empirical correlations related to average overtopping discharge were derived to analyze the resistance of levee against erosion. These equations could only fit to the situation similar to the experimental tests while they can give a reasonable estimation if applied to other conditions. Osouli et al. (2014) and Karimpour et al. (2015) conducted lab scale physical modeling of levees to evaluate score characterization of different levees due to floodwall overtopping.

Braided River

The Waimakariri River is braided over most of its course.

A braided river (or braided channel) is one of a number of channel types and has a channel that consists of a network of small channels separated by small and often temporary islands called braid bars or, in British usage, *aits* or *eyots*. Braided streams occur in rivers with high slope and/or large sediment load. Braided channels are also typical of environments that dramatically decrease channel depth, and consequently channel velocity, such as river deltas, alluvial fans and peneplains.

Formation

The White River in the U.S. state of Washington transports a large sediment load from the Emmons Glacier of Mount Rainier, a young, rapidly eroding volcano.

Braided rivers, as distinct from meandering rivers, occur when a threshold level of sediment load or slope is reached whilst a steep gradient is also maintained. Geologically speaking, an increase in sediment load will over time increase in the slope of the river, so these two conditions can be considered synonymous; and, consequently, a variation of slope can model a variation in sediment load. A threshold slope was experimentally determined to be 0.016 (ft/ft) for a 0.15 cu ft/s (0.0042 m³/s) stream with poorly sorted coarse sand. Any slope over this threshold created a braided stream, while any slope under the threshold created a meandering stream or— for very low slopes—a straight channel. So the main controlling factor on river development is the amount of sediment that the river carries; once a given system crosses a threshold value for sediment load, it will convert from a meandering system to a braided system. Also important to channel development is the proportion of suspended load sediment to bed load. An increase in suspended sediment allowed for the deposition of fine erosion-resistant material on the inside of a curve, which accentuated the curve and in some instances caused a river to shift from a braided to a meandering profile. The channels and braid bars are usually highly mobile, with the river layout often changing significantly during flood events. Channels move sideways via differential velocity: On the outside of a curve, deeper, swift water picks up sediment (usually gravel or larger stones), which is re-deposited in slow-moving water on the inside of a bend.

The braided channels may flow within an area defined by relatively stable banks or may occupy an entire valley floor. The Rakaia River in Canterbury, New Zealand has cut a channel 100 metres wide into the surrounding plains; this river transports sediment to a lagoon located on the river-coast interface.

Conditions associated with braided channel formation include:

- an abundant supply of sediment

- high stream gradient

- rapid and frequent variations in water discharge

- erodible banks

- a steep channel gradient

However, the critical factor that determines whether a stream will meander or braid is bank erodibility. A stream with cohesive banks that are resistant to erosion will form narrow, deep, meandering channels, whereas a stream with highly erodible banks will form wide, shallow channels, sustaining helical flow and resulting in the formation of braided channels.

Examples

Extensive braided river systems are found in Alaska, Canada, New Zealand's South Island, and the Himalayas, which all contain young, rapidly eroding mountains.

- The enormous Brahmaputra-Jamuna River in Asia is a classic example of a braided river.

- Braided river systems are present in Africa, for example in the Touat Valley.

- A notable example of a large braided stream in the contiguous United States is the Platte River in central and western Nebraska. The sediment of the arid Great Plains is augmented by the presence of the nearby Sandhills region north of the river.

- A portion of the lower Yellow River takes a braided form.

- The Sewanee Conglomerate, a Pennsylvanian coarse sandstone and conglomerate unit present on the Cumberland Plateau near the University of the South, may have been deposited by an ancient braided and meandering river that once existed in the eastern United States. Others have interpreted the depositional environment for this unit as a tidal delta.

Notable braided rivers in Europe:

- Italy

  - Tagliamento (Northeastern Italy)

  - Piave (river)

  - Brenta (river)

  - Cellina

  - Meduna

  - Fella

  - Magra

- Narew (Poland and Belarus)

Distinction Between Braided Rivers and Anastomosing Rivers

Anastomosing rivers or streams are similar to braided rivers in that they consist of multiple interweaving channels. However, anastomosing rivers typically consist of a network of low-gradient, narrow, deep channels with stable banks, in contrast to braided rivers, which form on steeper gradients and display less bank stability.

References

- Prothero, Donald R.; Schwab, Fred (1996), Sedimentary Geology: An Introduction to Sedimentary Rocks and Stratigraphy, W. H. Freeman, ISBN 0-7167-2726-9

- Reading, H. G. (1978), Sedimentary Environments: Processes, Facies and Stratigraphy, Cambridge, MA: Blackwell Science, ISBN 0-632-03627-3

- Klaus K.E. Neuendorf, James P. Mehl, Jr., Julia A. Jackson., eds. (2005). Glossary of geology. Alexandria: American Geological Institute. p. 382. ISBN 0-922152-76-4

- Renaud, F. and C. Kuenzer 2012: The Mekong Delta System - Interdisciplinary Analyses of a River Delta, Springer, ISBN 978-94-007-3961-1.

- Catling, David (1992). Rice in deep water. International Rice Research Institute. p. 177. ISBN 978-971-22-0005-2. Retrieved 23 April 2011.

- Bowker, Kent A. (1988). "Albert Einstein and Meandering Rivers". Earth Science History. 1 (1). Retrieved 2016-07-01.

- Harwood, William; Wall, Mike (September 27, 2012). "Mars rover Curiosity finds ancient stream bed". CBS News. Retrieved January 21, 2016.

- J. Radebaugh; et al. (2013). "Alluvial Fans on Titan Reveal Materials, Processes and Regional Conditions" (PDF). 44th Lunar and Planetary Science Conference. Retrieved January 21, 2016.

- "Hao Zhang, Hajime Nakagawa, 2008, ''Scour around Spur Dyke: Recent Advances and Future Researches''" (PDF). Retrieved 2013-05-17.

- "Weavers' Way footpath closure - Decoy Road (Hickling) to Potter Heigham 7 January 2011 - 6 April 2012". Countrysideaccess.norfolk.gov.uk. Retrieved 2013-05-17.

- "African dust caused red soil in southern Europe". Spanish Foundation for Science and Technology. November 11, 2010. Retrieved November 2, 2012.

3

Sedimentary Rocks and Structures

Sedimentary rocks are rocks that are formed by the deposition of Earth's surface. The sediment is in its initial stage formed by weathering and erosion from where it is transported to the place of deposition by water or wind. Sedimentary rocks can be classified into clastic rocks, phosphorite, iron-rich sedimentary rocks and carbonate rocks. The topics discussed in the section are of great importance to broaden the existing knowledge on sedimentary rocks.

Sedimentary Rock

Sedimentary rocks are types of rock that are formed by the deposition and subsequent cementation of that material at the Earth's surface and within bodies of water. Sedimentation is the collective name for processes that cause mineral and/or organic particles (detritus) to settle in place. The particles that form a sedimentary rock by accumulating are called sediment. Before being deposited, the sediment was formed by weathering and erosion from the source area, and then transported to the place of deposition by water, wind, ice, mass movement or glaciers, which are called agents of denudation. Sedimentation may also occur as minerals precipitate from water solution or shells of aquatic creatures settle out of suspension.

Middle Triassic marginal marine sequence of siltstones (reddish layers at the cliff base) and limestones (brown rocks above), Virgin Formation, southwestern Utah, USA

The sedimentary rock cover of the continents of the Earth's crust is extensive (73% of the Earth's current land surface), but the total contribution of sedimentary rocks is estimated to be only 8% of the total volume of the crust. Sedimentary rocks are only a thin veneer over a crust consisting mainly of igneous and metamorphic rocks. Sedimentary rocks are deposited in layers as strata, forming a structure called bedding. The study of sedimentary rocks and rock strata provides information about the subsurface that is useful for civil engineering, for example in the construction of roads, houses, tunnels, canals or other structures. Sedimentary rocks are also important sources of natural resources like coal, fossil fuels, drinking water or ores.

Sedimentary rocks on Mars, investigated by NASA's Curiosity Mars rover

Steeply dipping sedimentary rock strata along the Chalous Road in northern Iran

The study of the sequence of sedimentary rock strata is the main source for an understanding of the Earth's history, including palaeogeography, paleoclimatology and the history of life. The scientific discipline that studies the properties and origin of sedimentary rocks is called sedimentology. Sedimentology is part of both geology and physical geography and overlaps partly with other disciplines in the Earth sciences, such as pedology, geomorphology, geochemistry and structural geology. Sedimentary rocks have also been found on Mars.

Classi ication based on Origin

Sedimentary rocks can be subdivided into four groups based on the processes responsible for their formation: clastic sedimentary rocks, biochemical (biogenic) sedimentary rocks, chemical sedimentary rocks, and a fourth category for "other" sedimentary rocks formed by impacts, volcanism, and other minor processes.

Clastic Sedimentary Rocks

Clastic sedimentary rocks are composed of other rock fragments that were cemented by silicate minerals. Clastic rocks are composed largely of quartz, feldspar, rock (lithic)

fragments, clay minerals, and mica; any type of mineral may be present, but they in general represent the minerals that exist locally.

Clastic sedimentary rocks, are subdivided according to the dominant particle size. Most geologists use the Udden-Wentworth grain size scale and divide unconsolidated sediment into three fractions: gravel (>2 mm diameter), sand (1/16 to 2 mm diameter), and mud (clay is <1/256 mm and silt is between 1/16 and 1/256 mm). The classification of clastic sedimentary rocks parallels this scheme; conglomerates and breccias are made mostly of gravel, sandstones are made mostly of sand, and mudrocks are made mostly of the finest material. This tripartite subdivision is mirrored by the broad categories of rudites, arenites, and lutites, respectively, in older literature.

The subdivision of these three broad categories is based on differences in clast shape:- conglomerates and breccias), composition (sandstones), grain size and/or texture (mudrocks).

Conglomerates and Breccias

Conglomerates are dominantly composed of rounded gravel, while breccias are composed of dominantly angular gravel.

Sandstones

Sedimentary rock with sandstone in Malta

Sandstone classification schemes vary widely, but most geologists have adopted the Dott scheme, which uses the relative abundance of quartz, feldspar, and lithic framework grains and the abundance of a muddy matrix between the larger grains.

Composition of framework grains

The relative abundance of sand-sized framework grains determines the first word in a sandstone name. Naming depends on the dominance of the three most abundant components quartz, feldspar, or the lithic fragments that originated from other rocks. All other minerals are considered accessories and not used in the naming of the rock, regardless of abundance.

- Quartz sandstones have >90% quartz grains

- Feldspathic sandstones have <90% quartz grains and more feldspar grains than lithic grains

- Lithic sandstones have <90% quartz grains and more lithic grains than feldspar grains

Abundance of muddy matrix material between sand grains

When sand-sized particles are deposited, the space between the grains either remains open or is filled with mud (silt and/or clay sized particle).

- "Clean" sandstones with open pore space (that may later be filled with matrix material) are called arenites.

- Muddy sandstones with abundant (>10%) muddy matrix are called wackes.

Six sandstone names are possible using the descriptors for grain composition (quartz-, feldspathic-, and lithic-) and the amount of matrix (wacke or arenite). For example, a quartz arenite would be composed of mostly (>90%) quartz grains and have little or no clayey matrix between the grains, a lithic wacke would have abundant lithic grains and abundant muddy matrix, etc.

Although the Dott classification scheme is widely used by sedimentologists, common names like greywacke, arkose, and quartz sandstone are still widely used by non-specialists and in popular literature.

Mudrocks

Lower Antelope Canyon was carved out of the surrounding sandstone by both mechanical weathering and chemical weathering. Wind, sand, and water from flash flooding are the primary weathering agents.

Mudrocks are sedimentary rocks composed of at least 50% silt- and clay-sized particles. These relatively fine-grained particles are commonly transported by turbulent flow in water or air, and deposited as the flow calms and the particles settle out of suspension.

Most authors presently use the term "mudrock" to refer to all rocks composed dominantly of mud. Mudrocks can be divided into siltstones, composed dominantly of silt-sized particles; mudstones with subequal mixture of silt- and clay-sized particles; and claystones, composed mostly of clay-sized particles. Most authors use "shale" as a term for a fissile mudrock (regardless of grain size) although some older literature uses the term "shale" as a synonym for mudrock.

Biochemical Sedimentary Rocks

Outcrop of Ordovician oil shale (kukersite), northern Estonia

Biochemical sedimentary rocks are created when organisms use materials dissolved in air or water to build their tissue. Examples include:

- Most types of limestone are formed from the calcareous skeletons of organisms such as corals, mollusks, and foraminifera.

- Coal, formed from plants that have removed carbon from the atmosphere and combined it with other elements to build their tissue.

- Deposits of chert formed from the accumulation of siliceous skeletons of microscopic organisms such as radiolaria and diatoms.

Chemical Sedimentary Rocks

Chemical sedimentary rock forms when mineral constituents in solution become supersaturated and inorganically precipitate. Common chemical sedimentary rocks include oolitic limestone and rocks composed of evaporite minerals, such as halite (rock salt), sylvite, barite and gypsum.

"Other" Sedimentary Rocks

This fourth miscellaneous category includes rocks formed by Pyroclastic flows, impact breccias, volcanic breccias, and other relatively uncommon processes.

Compositional Classification Schemes

Alternatively, sedimentary rocks can be subdivided into compositional groups based on their mineralogy:

- Siliciclastic sedimentary rocks, are dominantly composed of silicate minerals. The sediment that makes up these rocks was transported as bed load, suspended load, or by sediment gravity flows. Siliciclastic sedimentary rocks are subdivided into conglomerates and breccias, sandstone, and mudrocks.

- Carbonate sedimentary rocks are composed of calcite (rhombohedral $CaCO_3$), aragonite (orthorhombic $CaCO_3$), dolomite ($CaMg(CO_3)_2$), and other carbonate minerals based on the CO_{2-3} ion. Common examples include limestone and dolostone.

- Evaporite sedimentary rocks are composed of minerals formed from the evaporation of water. The most common evaporite minerals are carbonates (calcite and others based on CO_{2-3}), chlorides (halite and others built on Cl^-), and sulfates (gypsum and others built on SO_{2-4}). Evaporite rocks commonly include abundant halite (rock salt), gypsum, and anhydrite.

- Organic-rich sedimentary rocks have significant amounts of organic material, generally in excess of 3% total organic carbon. Common examples include coal, oil shale as well as source rocks for oil and natural gas.

- Siliceous sedimentary rocks are almost entirely composed of silica (SiO_2), typically as chert, opal, chalcedony or other microcrystalline forms.

- Iron-rich sedimentary rocks are composed of >15% iron; the most common forms are banded iron formations and ironstones.

- Phosphatic sedimentary rocks are composed of phosphate minerals and contain more than 6.5% phosphorus; examples include deposits of phosphate nodules, bone beds, and phosphatic mudrocks.

Deposition and Transformation

Sediment Transport and Deposition

Sedimentary rocks are formed when sediment is deposited out of air, ice, wind, gravity, or water flows carrying the particles in suspension. This sediment is often formed when weathering and erosion break down a rock into loose material in a source area. The material is then transported from the source area to the deposition area. The type of sediment transported depends on the geology of the hinterland (the source area of the sediment). However, some sedimentary rocks, such as evaporites, are composed of material that form at the place of deposition. The nature of a sedimentary rock, therefore, not only depends on the sediment supply, but also on the sedimentary depositional environment in which it formed.

Cross-bedding and scour in a fine sandstone; the Logan Formation (Mississippian) of Jackson County, Ohio

Transformation (Diagenesis)

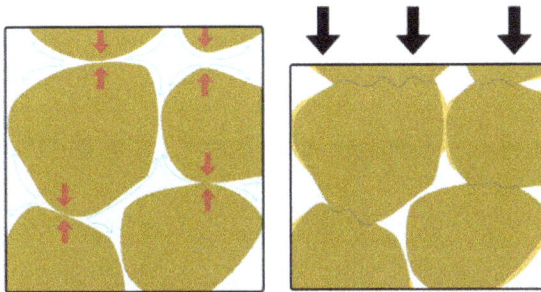

Pressure solution at work in a clastic rock. While material dissolves at places where grains are in contact, that material may recrystallize from the solution and act as cement in open pore spaces. As a result, there is a net flow of material from areas under high stress to those under low stress, producing a sedimentary rock becomes more compact and harder. Loose sand can become sandstone in this way.

The term diagenesis is used to describe all the chemical, physical, and biological changes, exclusive of surface weathering, undergone by a sediment after its initial deposition. Some of those processes cause the sediment to consolidate into a compact, solid substance from the originally loose material. Young sedimentary rocks, especially those of Quaternary age (the most recent period of the geologic time scale) are often still unconsolidated. As sediment deposition builds up, the overburden (lithostatic) pressure rises, and a process known as lithification takes place.

Sedimentary rocks are often saturated with seawater or groundwater, in which minerals can dissolve, or from which minerals can precipitate. Precipitating minerals reduce the pore space in a rock, a process called cementation. Due to the decrease in pore space, the original connate fluids are expelled. The precipitated minerals form a cement and make the rock more compact and competent. In this way, loose clasts in a sedimentary rock can become "glued" together.

When sedimentation continues, an older rock layer becomes buried deeper as a result. The lithostatic pressure in the rock increases due to the weight of the overlying sediment. This causes compaction, a process in which grains mechanically reorganize. Compaction is, for example, an important diagenetic process in clay, which can initially consist of 60% water. During compaction, this interstitial water is pressed out of pore spaces. Compaction can also be the result of dissolution of grains by pressure solution. The dissolved material precipitates again in open pore spaces, which means there is a net flow of material into the pores. However, in some cases, a certain mineral dissolves and does not precipitate again. This process, called leaching, increases pore space in the rock.

Some biochemical processes, like the activity of bacteria, can affect minerals in a rock and are therefore seen as part of diagenesis. Fungi and plants (by their roots) and various other organisms that live beneath the surface can also influence diagenesis.

Burial of rocks due to ongoing sedimentation leads to increased pressure and temperature, which stimulates certain chemical reactions. An example is the reactions by which organic material becomes lignite or coal. When temperature and pressure increase still further, the realm of diagenesis makes way for metamorphism, the process that forms metamorphic rock.

Properties

A piece of a banded iron formation, a type of rock that consists of alternating layers with iron(III) oxide (red) and iron(II) oxide (grey). BIFs were mostly formed during the Precambrian, when the atmosphere was not yet rich in oxygen. Moories Group, Barberton Greenstone Belt, South Africa

Color

The color of a sedimentary rock is often mostly determined by iron, an element with two major oxides: iron(II) oxide and iron(III) oxide. Iron(II) oxide (FeO) only forms under low oxygen (anoxic) circumstances and gives the rock a grey or greenish colour. Iron(III) oxide (Fe_2O_3) in a richer iron environment is often found in the form of the mineral hematite and gives the rock a reddish to brownish colour. In arid continental climates rocks are in direct contact with the atmosphere, and oxidation is an important process, giving the rock a red or orange colour. Thick sequences of red sedimentary rocks formed in arid climates are called red beds. However, a red colour does not necessarily mean the rock formed in a continental environment or arid climate.

The presence of organic material can colour a rock black or grey. Organic material is formed from dead organisms, mostly plants. Normally, such material eventually decays by oxidation or bacterial activity. Under anoxic circumstances, however, organic material cannot decay and leaves a dark sediment, rich in organic material. This can, for example, occur at the bottom of deep seas and lakes. There is little water mixing in such environments, as a result oxygen from surface water is not brought down, and the deposited sediment is normally a fine dark clay. Dark rocks, rich in organic material, are therefore often shales.

Texture

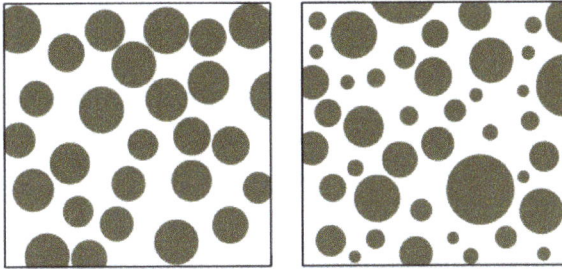

Diagram showing well-sorted (left) and poorly sorted (right) grains

The size, form and orientation of clasts (the original pieces of rock) in a sediment is called its texture. The texture is a small-scale property of a rock, but determines many of its large-scale properties, such as the density, porosity or permeability.

The 3D orientation of the clasts is called the fabric of the rock. Between the clasts, the rock can be composed of a matrix (a cement) that consists of crystals of one or more precipitated minerals. The size and form of clasts can be used to determine the velocity and direction of current in the sedimentary environment that moved the clasts from their origin; fine, calcareous mud only settles in quiet water while gravel and larger clasts are moved only by rapidly moving water. The grain size of a rock is usually expressed with the Wentworth scale, though alternative scales are sometimes used. The grain size can be expressed as a diameter or a volume, and is always an average value – a rock is composed of clasts with different sizes. The statistical distribution of grain sizes is different for different rock types and is described in a property called the sorting of the rock. When all clasts are more or less of the same size, the rock is called 'well-sorted', and when there is a large spread in grain size, the rock is called 'poorly sorted'.

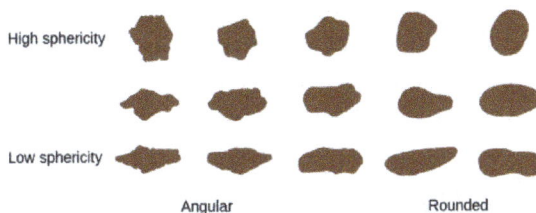

High sphericity

Low sphericity

Angular   Rounded

Diagram showing the rounding and sphericity of grains

The form of the clasts can reflect the origin of the rock.

Coquina, a rock composed of clasts of broken shells, can only form in energetic water. The form of a clast can be described by using four parameters:

- *Surface texture* describes the amount of small-scale relief of the surface of a grain that is too small to influence the general shape.

- *rounding* describes the general smoothness of the shape of a grain.

- 'Sphericity' describes the degree to which the grain approaches a sphere.

- 'Grain form' describes the three dimensional shape of the grain.

Chemical sedimentary rocks have a non-clastic texture, consisting entirely of crystals. To describe such a texture, only the average size of the crystals and the fabric are necessary.

Mineralogy

Most sedimentary rocks contain either quartz (especially siliciclastic rocks) or calcite (especially carbonate rocks). In contrast to igneous and metamorphic rocks, a sedimentary rock usually contains very few different major minerals. However, the origin of the minerals in a sedimentary rock is often more complex than in an igneous rock. Minerals in a sedimentary rock can have formed by precipitation during sedimentation or by diagenesis. In the second case, the mineral precipitate can have grown over an older generation of cement. A complex diagenetic history can be studied by optical mineralogy, using a petrographic microscope.

Carbonate rocks dominantly consist of carbonate minerals such as calcite, aragonite or dolomite. Both the cement and the clasts (including fossils and ooids) of a carbonate sedimentary rock can consist of carbonate minerals. The mineralogy of a clastic rock is determined by the material supplied by the source area, the manner of its transport to the place of deposition and the stability of that particular mineral. The resistance of rock forming minerals to weathering is expressed by Bowen's reaction series. In this series, quartz is the most stable, followed by feldspar, micas, and finally other less stable minerals that are only present when little weathering has occurred. The amount of weathering depends mainly on the distance to the source area, the local climate and the time it took for the sediment to be transported to the point where it is deposited. In most sedimentary rocks, mica, feldspar and less stable minerals have been reduced to clay minerals like kaolinite, illite or smectite.

Fossils

Among the three major types of rock, fossils are most commonly found in sedimentary rock. Unlike most igneous and metamorphic rocks, sedimentary rocks form at temperatures and pressures that do not destroy fossil remnants. Often these fossils may only be visible under magnification.

Fossil-rich layers in a sedimentary rock, Año Nuevo State Reserve, California

Dead organisms in nature are usually quickly removed by scavengers, bacteria, rotting and erosion, but sedimentation can contribute to exceptional circumstances where these natural processes are unable to work, causing fossilisation. The chance of fossilisation is higher when the sedimentation rate is high (so that a carcass is quickly buried), in anoxic environments (where little bacterial activity occurs) or when the organism had a particularly hard skeleton. Larger, well-preserved fossils are relatively rare.

Burrows in a turbidite, made by crustaceans, San Vincente Formation (early Eocene) of the Ainsa Basin, southern foreland of the Pyrenees

Fossils can be both the direct remains or imprints of organisms and their skeletons. Most commonly preserved are the harder parts of organisms such as bones, shells, and the woody tissue of plants. Soft tissue has a much smaller chance of being fossilized, and the preservation of soft tissue of animals older than 40 million years is very rare. Imprints of organisms made while they were still alive are called trace fossils, examples of which are burrows, footprints, etc.

As a part of a sedimentary or metamorphic rock, fossils undergo the same diagenetic processes as does the containing rock. A shell consisting of calcite can, for example, dissolve while a cement of silica then fills the cavity. In the same way, precipitating minerals can fill cavities formerly occupied by blood vessels, vascular tissue or other soft tissues. This preserves the form of the organism but changes the chemical composition, a process called permineralization. The most common minerals involved in permineralization are cements of carbonates (especially calcite), forms of amorphous silica (chalcedony, flint, chert) and pyrite. In the case of silica cements, the process is called lithification.

At high pressure and temperature, the organic material of a dead organism undergoes chemical reactions in which volatiles such as water and carbon dioxide are expulsed. The fossil, in the end, consists of a thin layer of pure carbon or its mineralized form, graphite. This form of fossilisation is called carbonisation. It is particularly important for plant fossils. The same process is responsible for the formation of fossil fuels like lignite or coal.

Primary Sedimentary Structures

Cross-bedding in a fluviatile sandstone, Middle Old Red Sandstone (Devonian) on Bressay, Shetland Islands

A flute cast, a type of sole marking, from the Book Cliffs of Utah

Structures in sedimentary rocks can be divided into 'primary' structures (formed during deposition) and 'secondary' structures (formed after deposition). Unlike textures, structures are always large-scale features that can easily be studied in the field. Sedimentary structures can indicate something about the sedimentary environment or can serve to tell which side originally faced up where tectonics have tilted or overturned sedimentary layers.

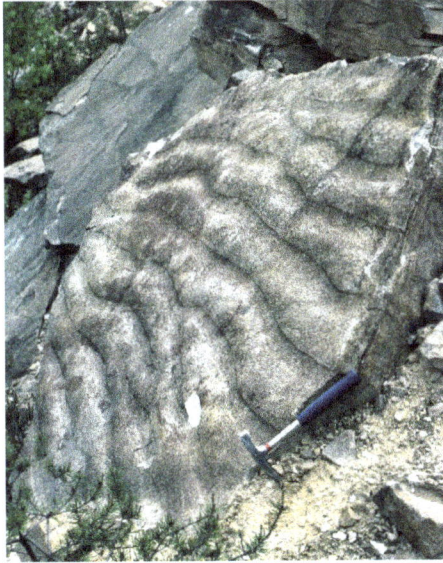

Ripple marks formed by a current in a sandstone that was later tilted (Haßberge, Bavaria)

Sedimentary rocks are laid down in layers called beds or strata. A bed is defined as a layer of rock that has a uniform lithology and texture. Beds form by the deposition of layers of sediment on top of each other. The sequence of beds that characterizes sedimentary rocks is called bedding. Single beds can be a couple of centimetres to several meters thick. Finer, less pronounced layers are called laminae, and the structure it forms in a rock is called lamination. Laminae are usually less than a few centimetres thick. Though bedding and lamination are often originally horizontal in nature, this is not always the case. In some environments, beds are deposited at a (usually small) angle. Sometimes multiple sets of layers with different orientations exist in the same rock, a structure called cross-bedding. Cross-bedding forms when small-scale erosion occurs during deposition, cutting off part of the beds. Newer beds then form at an angle to older ones.

The opposite of cross-bedding is parallel lamination, where all sedimentary layering is parallel. Differences in laminations are generally caused by cyclic changes in the sediment supply, caused, for example, by seasonal changes in rainfall, temperature or biochemical activity. Laminae that represent seasonal changes (similar to tree rings) are called varves. Any sedimentary rock composed of millimeter or finer scale layers can be named with the general term *laminite*. When sedimentary rocks have no lamination at all, their structural character is called massive bedding.

Graded bedding is a structure where beds with a smaller grain size occur on top of beds with larger grains. This structure forms when fast flowing water stops flowing. Larger, heavier clasts in suspension settle first, then smaller clasts. Although graded bedding can form in many different environments, it is a characteristic of turbidity currents.

The surface of a particular bed, called the bedform, can be indicative of a particular sedimentary environment, too. Examples of bed forms include dunes and ripple marks. Sole markings, such as tool marks and flute casts, are groves dug into a sedimentary layer that are preserved. These are often elongated structures and can be used to establish the direction of the flow during deposition.

Ripple marks also form in flowing water. There are two types of ripples: symmetric and asymmetric. Environments where the current is in one direction, such as rivers, produce asymmetric ripples. The longer flank of such ripples is on the upstream side of the current. Symmetric wave ripples occur in environments where currents reverse directions, such as tidal flats.

Mudcracks are a bed form caused by the dehydration of sediment that occasionally comes above the water surface. Such structures are commonly found at tidal flats or point bars along rivers.

Secondary Sedimentary Structures

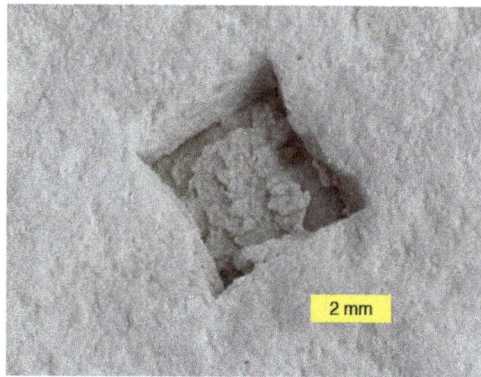

Halite crystal mold in dolomite, Paadla Formation (Silurian), Saaremaa, Estonia

Secondary sedimentary structures are those which formed after deposition. Such structures form by chemical, physical and biological processes within the sediment. They can be indicators of circumstances after deposition. Some can be used as way up criteria.

Organic materials in a sediment can leave more traces than just fossils. Preserved tracks and burrows are examples of trace fossils (also called ichnofossils). Such traces are relatively rare. Most trace fossils are burrows of molluscs or arthropods. This burrowing is called bioturbation by sedimentologists. It can be a valuable indicator of the biological and ecological environment that existed after the sediment was deposited. On the other hand, the burrowing activity of organisms can destroy other (primary) structures in the sediment, making a reconstruction more difficult.

Secondary structures can also form by diagenesis or the formation of a soil (pedogenesis) when a sediment is exposed above the water level. An example of a diagenetic structure common in carbonate rocks is a stylolite. Stylolites are irregular planes where material was dissolved into the pore fluids in the rock. This can result in the precipitation of a certain chemical species producing colouring and staining of the rock, or the formation of concretions. Concretions are roughly concentric bodies with a different composition from the host rock. Their formation can be the result of localized precipitation due to small differences in composition or porosity of the host rock, such as around fossils, inside burrows or around plant roots. In carbonate based rocks such as limestone or chalk, chert or flint concretions are common, while terrestrial sandstones can have iron concretions. Calcite concretions in clay are called septarian concretions.

Chert concretions in chalk, Middle Lefkara Formation (upper Paleocene to middle Eocene), Cyprus

After deposition, physical processes can deform the sediment, producing a third class of secondary structures. Density contrasts between different sedimentary layers, such as between sand and clay, can result in flame structures or load casts, formed by inverted diapirism. While the clastic bed is still fluid, diapirism can cause a denser upper layer to sink into a lower layer. Sometimes, density contrasts can result or grow when one of the lithologies dehydrates. Clay can be easily compressed as a result of dehydration, while sand retains the same volume and becomes relatively less dense. On the other hand, when the pore fluid pressure in a sand layer surpasses a critical point, the sand can break through overlying clay layers and flow through, forming discordant bodies of sedimentary rock called sedimentary dykes. The same process can form mud volcanoes on the surface where they broke through upper layers.

Sedimentary dykes can also be formed in a cold climate where the soil is permanently frozen during a large part of the year. Frost weathering can form cracks in the soil that fill with rubble from above. Such structures can be used as climate indicators as well as way up structures.

Density contrasts can also cause small-scale faulting, even while sedimentation progresses (synchronous-sedimentary faulting). Such faulting can also occur when large masses of non-lithified sediment are deposited on a slope, such as at the front side of a delta or the continental slope. Instabilities in such sediments can result in the deposited material to slump, producing fissures and folding. The resulting structures in the rock are syn-sedimentary folds and faults, which can be difficult to distinguish from folds and faults formed by tectonic forces acting on lithified rocks.

Sedimentary Environments

The setting in which a sedimentary rock forms is called the sedimentary environment. Every environment has a characteristic combination of geologic processes and circumstances. The type of sediment that is deposited is not only dependent on the sediment that is transported to a place, but also on the environment itself.

A marine environment means that the rock was formed in a sea or ocean. Often, a distinction is made between deep and shallow marine environments. Deep marine usually refers to environments more than 200 m below the water surface. Shallow marine environments exist adjacent to coastlines and can extend to the boundaries of the continental shelf. The water movements in such environments have a generally higher energy than that in deep environments, as wave activity diminishes with depth. This means that coarser sediment particles can be transported and the deposited sediment can be coarser than in deeper environments. When the sediment is transported from the continent, an alternation of sand, clay and silt is deposited. When the continent is far away, the amount of such sediment deposited may be small, and biochemical processes dominate the type of rock that forms. Especially in warm climates, shallow marine environments far offshore mainly see deposition of carbonate rocks. The shallow, warm water is an ideal habitat for many small organisms that build carbonate skeletons. When these organisms die, their skeletons sink to the bottom, forming a thick layer of calcareous mud that may lithify into limestone. Warm shallow marine environments also are ideal environments for coral reefs, where the sediment consists mainly of the calcareous skeletons of larger organisms.

In deep marine environments, the water current working the sea bottom is small. Only fine particles can be transported to such places. Typically sediments depositing on the ocean floor are fine clay or small skeletons of micro-organisms. At 4 km depth, the solubility of carbonates increases dramatically (the depth zone where this happens is called the lysocline). Calcareous sediment that sinks below the lysocline dissolves, as a result no limestone can be formed below this depth. Skeletons of micro-organisms formed of silica (such as radiolarians) are not as soluble and still deposit. An example of a rock formed of silica skeletons is radiolarite. When the bottom of the sea has a small inclination, for example at the continental slopes, the sedimentary cover can become unstable, causing turbidity currents. Turbidity currents are sudden disturbances of the normally quite deep marine environment and can cause the geologically speaking instantaneous deposition of large amounts of sediment, such as sand and silt. The rock sequence formed by a turbidity current is called a turbidite.

The coast is an environment dominated by wave action. At a beach, dominantly denser sediment such as sand or gravel, often mingled with shell fragments, is deposited, while the silt and clay sized material is kept in mechanical suspension. Tidal flats and shoals are places that sometimes dry because of the tide. They are often cross-cut by gullies, where the current is strong and the grain size of the deposited sediment is larger. Where rivers enter the body of water, either on a sea or lake coast, deltas can form. These are large accumulations of sediment transported from the continent to places in front of the mouth of the river. Deltas are dominantly composed of clastic sediment (in contrast to chemical).

A sedimentary rock formed on land has a continental sedimentary environment. Examples of continental environments are lagoons, lakes, swamps, floodplains and alluvial fans. In the quiet water of swamps, lakes and lagoons, fine sediment is deposited, mingled with organic material from dead plants and animals. In rivers, the energy of the water is much greater and can transport heavier clastic material. Besides transport by water, sediment can in continental environments also be transported by wind or glaciers. Sediment transported by wind is called aeolian and is always very well sorted, while sediment transported by a glacier is called glacial till and is characterized by very poor sorting.

Aeolian deposits can be quite striking. The depositional environment of the Touchet Formation, located in the Northwestern United States, had intervening periods of aridity which resulted in a series of rhythmite layers. Erosional cracks were later infilled with layers of soil material, especially from aeolian processes. The infilled sections formed vertical inclusions in the horizontally deposited layers of the Touchet Formation, and thus provided evidence of the events that intervened over time among the forty-one layers that were deposited.

Sedimentary Facies

Sedimentary environments usually exist alongside each other in certain natural successions. A beach, where sand and gravel is deposited, is usually bounded by a deeper marine environment a little offshore, where finer sediments are deposited at the same time. Behind the beach, there can be dunes (where the dominant deposition is well sorted sand) or a lagoon (where fine clay and organic material is deposited). Every sedimentary environment has its own characteristic deposits. The typical rock formed in a certain environment is called its sedimentary facies. When sedimentary strata accumulate through time, the environment can shift, forming a change in facies in the subsurface at one location. On the other hand, when a rock layer with a certain age is followed laterally, the lithology (the type of rock) and facies eventually change.

Facies can be distinguished in a number of ways: the most common are by the lithology (for example: limestone, siltstone or sandstone) or by fossil content. Coral for example only lives in warm and shallow marine environments and fossils of coral are thus typical for shallow marine facies. Facies determined by lithology are called lithofacies; facies determined by fossils are biofacies.

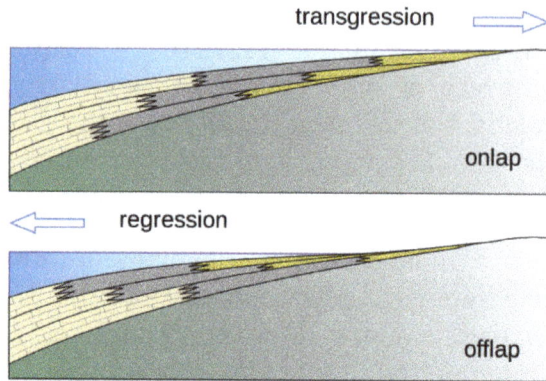

Shifting sedimentary facies in the case of transgression (above) and regression of the sea (below)

Sedimentary environments can shift their geographical positions through time. Coastlines can shift in the direction of the sea when the sea level drops, when the surface rises due to tectonic forces in the Earth's crust or when a river forms a large delta. In the subsurface, such geographic shifts of sedimentary environments of the past are recorded in shifts in sedimentary facies. This means that sedimentary facies can change either parallel or perpendicular to an imaginary layer of rock with a fixed age, a phenomenon described by Walther's Law.

The situation in which coastlines move in the direction of the continent is called transgression. In the case of transgression, deeper marine facies are deposited over shallower facies, a succession called onlap. Regression is the situation in which a coastline moves in the direction of the sea. With regression, shallower facies are deposited on top of deeper facies, a situation called offlap.

The facies of all rocks of a certain age can be plotted on a map to give an overview of the palaeogeography. A sequence of maps for different ages can give an insight in the development of the regional geography.

Sedimentary Basins

Places where large-scale sedimentation takes place are called sedimentary basins. The amount of sediment that can be deposited in a basin depends on the depth of the basin, the so-called accommodation space. The depth, shape and size of a basin depend on tectonics, movements within the Earth's lithosphere. Where the lithosphere moves upward (tectonic uplift), land eventually rises above sea level, so that and erosion removes material, and the area becomes a source for new sediment. Where the lithosphere moves downward (tectonic subsidence), a basin forms and sedimentation can take place. When the lithosphere keeps subsiding, new accommodation space keeps being created.

A type of basin formed by the moving apart of two pieces of a continent is called a rift basin. Rift basins are elongated, narrow and deep basins. Due to divergent movement, the lithosphere is stretched and thinned, so that the hot asthenosphere rises and heats the overlying rift basin. Apart from continental sediments, rift basins normally also have part of their infill consisting of volcanic deposits. When the basin grows due to continued stretching of the lithosphere, the rift grows and the sea can enter, forming marine deposits.

When a piece of lithosphere that was heated and stretched cools again, its density rises, causing isostatic subsidence. If this subsidence continues long enough, the basin is called a sag basin. Examples of sag basins are the regions along passive continental margins, but sag basins can also be found in the interior of continents. In sag basins, the extra weight of the newly deposited sediments is enough to keep the subsidence going in a vicious circle. The total thickness of the sedimentary infill in a sag basins can thus exceed 10 km.

A third type of basin exists along convergent plate boundaries - places where one tectonic plate moves under another into the asthenosphere. The subducting plate bends and forms a fore-arc basin in front of the overriding plate—an elongated, deep asymmetric basin. Fore-arc basins are filled with deep marine deposits and thick sequences of turbidites. Such infill is called flysch. When the convergent movement of the two plates results in continental collision, the basin becomes shallower and develops into a foreland basin. At the same time, tectonic uplift forms a mountain belt in the overriding plate, from which large amounts of material are eroded and transported to the basin. Such erosional material of a growing mountain chain is called molasse and has either a shallow marine or a continental facies.

At the same time, the growing weight of the mountain belt can cause isostatic subsidence in the area of the overriding plate on the other side to the mountain belt. The basin type resulting from this subsidence is called a back-arc basin and is usually filled by shallow marine deposits and molasse.

Cyclic alternation of competent and less competent beds in the Blue Lias at Lyme Regis, southern England

Influence of Astronomical Cycles

In many cases facies changes and other lithological features in sequences of sedimentary rock have a cyclic nature. This cyclic nature was caused by cyclic changes in sediment supply and the sedimentary environment. Most of these cyclic changes are caused by astronomic cycles. Short astronomic cycles can be the difference between the tides or the spring tide every two weeks. On a larger time-scale, cyclic changes in climate and sea level are caused by Milankovitch cycles: cyclic changes in the orientation and/or position of the Earth's rotational axis and orbit around the Sun. There are a number of Milankovitch cycles known, lasting between 10,000 and 200,000 years.

Relatively small changes in the orientation of the Earth's axis or length of the seasons can be a major influence on the Earth's climate. An example are the ice ages of the past 2.6 million years (the Quaternary period), which are assumed to have been caused by astronomic cycles. Climate change can influence the global sea level (and thus the amount of accommodation space in sedimentary basins) and sediment supply from a certain region. Eventually, small changes in astronomic parameters can cause large changes in sedimentary environment and sedimentation.

Sedimentation Rates

The rate at which sediment is deposited differs depending on the location. A channel in a tidal flat can see the deposition of a few metres of sediment in one day, while on the deep ocean floor each year only a few millimetres of sediment accumulate. A distinction can be made between normal sedimentation and sedimentation caused by catastrophic processes. The latter category includes all kinds of sudden exceptional processes like mass movements, rock slides or flooding. Catastrophic processes can see the sudden deposition of a large amount of sediment at once. In some sedimentary environments, most of the total column of sedimentary rock was formed by catastrophic processes, even though the environment is usually a quiet place. Other sedimentary environments are dominated by normal, ongoing sedimentation.

In many cases, sedimentation occurs slowly. In a desert, for example, the wind deposits siliciclastic material (sand or silt) in some spots, or catastrophic flooding of a wadi may cause sudden deposits of large quantities of detrital material, but in most places eolian erosion dominates. The amount of sedimentary rock that forms is not only dependent on the amount of supplied material, but also on how well the material consolidates. Erosion removes most deposited sediment shortly after deposition.

Stratigraphy

The Permian through Jurassic stratigraphy of the Colorado Plateau area of southeastern Utah that makes up much of the famous prominent rock formations in protected areas such as Capitol Reef National Park and Canyonlands National Park. From top to bottom: Rounded tan domes of the Navajo Sandstone, layered red Kayenta Formation, cliff-forming, vertically jointed, red Wingate Sandstone, slope-forming, purplish Chinle Formation, layered, lighter-red Moenkopi Formation, and white, layered Cutler Formation sandstone. Picture from Glen Canyon National Recreation Area, Utah.

That new rock layers are above older rock layers is stated in the principle of superposition. There are usually some gaps in the sequence called unconformities. These represent periods where no new sediments were laid down, or when earlier sedimentary layers were raised above sea level and eroded away.

Sedimentary rocks contain important information about the history of the Earth. They contain fossils, the preserved remains of ancient plants and animals. Coal is considered a type of sedimentary rock. The composition of sediments provides us with clues as to the original rock. Differences between successive layers indicate changes to the environment over time. Sedimentary rocks can contain fossils because, unlike most igneous and metamorphic rocks, they form at temperatures and pressures that do not destroy fossil remains.

Classification of Sedimentary Rocks

Clastic Rock

A thin section of a clast (sand grain), derived from a basalt scoria. Vesicles (air bubbles) can be seen throughout the clast. Plane light above, cross-polarized light below. Scale box is 0.25 mm

Clastic rocks are composed of fragments, or clasts, of pre-existing minerals and rock. A clast is a fragment of geological detritus, chunks and smaller grains of rock broken off other rocks by physical weathering. Geologists use the term clastic with reference to sedimentary rocks as well as to particles in sediment transport whether in suspension or as bed load, and in sediment deposits.

Sedimentary Clastic Rocks

Clastic sedimentary rocks are rocks composed predominantly of broken pieces or *clasts* of older weathered and eroded rocks. Clastic sediments or sedimentary rocks are classified based on grain size, clast and cementing material (matrix) composition, and texture. The classification factors are often useful in determining a sample's environment of deposition. An example clastic environment would be a river system in which the full range of grains being transported by the moving water consist of pieces eroded from solid rock upstream.

Grain size varies from clay in shales and claystones; through silt in siltstones; sand in sandstones; and gravel, cobble, to boulder sized fragments in conglomerates and breccias. The Krumbein phi (φ) scale numerically orders these terms in a logarithmic size scale.

Siliciclastic Sedimentary Rocks

The term siliciclastic pertains to clastic noncarbonate rocks which are almost exclusively silicon-bearing, either as forms of quartz or silicates.

Composition

The composition of siliciclastic sedimentary rocks includes the chemical and mineralogical components of the framework as well as the cementing material that make up these rocks. Boggs divides them into four categories; major minerals, accessory minerals, rock fragments, and chemical sediments.

Major minerals can be categorized into subdivisions based on their resistance to chemical decomposition. Those that possess a great resistance to decomposition are categorized as stable, while those that do not are considered less stable. The most common stable mineral in siliciclastic sedimentary rocks is quartz. Quartz makes up approximately 65 percent of framework grains present in sandstones and about 30 percent of minerals in the average shale. Less stable minerals present in this type of rocks are feldspars, including both potassium and plagioclase feldspars. Feldspars comprise a considerably lesser portion of framework grains and minerals. They only make up about 15 percent of framework grains in sandstones and 5% of minerals in shales. Clay mineral groups are mostly present in mudrocks (comprising more than 60% of the minerals) but can be found in other siliciclastic sedimentary rocks at considerably lower levels.

Accessory minerals are associated with those whose presence in the rock are not directly important to the classification of the specimen. These generally occur in smaller amounts in comparison to the quartz, and feldspars. Furthermore, those that do occur are generally heavy minerals or coarse grained micas (both Muscovite and Biotite).

Rock fragments also occur in the composition of siliciclastic sedimentary rocks and are responsible for about 10 - 15 percent of the composition of sandstone. They generally make up most of the gravel size particles in conglomerates but contribute only a very small amount to the composition of mudrocks. Though they sometimes are, rock fragments are not always sedimentary in origin. They can also be metamorphic or igneous.

Chemical cements vary in abundance but are predominantly found in sandstones. The two major types, are silicate based and carbonate based. The majority of silica cements are composed of quartz but can include, chert, opal, feldspars and zeolites.

Composition includes the chemical and mineralogic make-up of the single or varied fragments and the cementing material (matrix) holding the clasts together as a rock. These differences are most commonly used in the framework grains of sandstones. Sandstones rich in quartz are called quartz arenites, those rich in feldspar are called arkoses, and those rich in lithics are called lithic sandstones.

Classification

Clastic Sedimentary Rocks

Texture (grain size)	Sediment Name	Rock Name
Coarse (over 2 mm)	Gravel (rounded fragments)	Conglomerate
	Gravel (angular fragments)	Breccia
Medium (1/16 to 2 mm)	Sand	Sandstone
Fine (1/16 to 1/256 mm)	Mud	Siltstone
Very Fine (less than 1/256)	Mud	Shale

Chemical Sedimentary Rocks

Composition	Texture (grain size)	Rock Name	
Calcite	Fine to coarse crystalline	Crystalline Limestone	
		Travertine	
	Shells and cemented shell fragments	Coquina	
	Shells and shell fragments cemented with calcite cement	Fossiliferous Limestone	Biochemical Limestone
	Microscopic shells and clay	Chalk	
Quartz	Very fine crystalline	Chert (light color) Flint (dark color)	
Gypsum	Fine to coarse crystalline	Rock Gypsum	
Halite	Fine to coarse crystalline	Rock Salt	
Altered plant fragments	Fine-grained organic matter	Bituminous Coal	

Sedimentary Rock Chart

Siliciclastic sedimentary rocks are composed of mainly silicate particles derived by the weathering of older rocks and pyroclastic volcanism. While grain size, clast and cementing material (matrix) composition, and texture are important factors when regarding composition, siliciclastic sedimentary rocks are classified according to grain size into three major categories; conglomerates, sandstones, and mudrocks. The term clay is used to classify particles smaller than .0039 millimeters. However, term can also be used to refer to a family of sheet silicate minerals. Silt refers to particles that have a diameter between .062 and .0039 millimeters. The term mud is used to refer to when clay and silt particles often mix to create aggregates sediments. Hence, the term *mudrock* is used to refer to rocks that are composed of silt and clay particles. Furthermore, particles that reach diameters between .062 and 2 millimeters fall into the category of sand. When sand is cemented together and lithified it becomes known as sandstone. Any particle that is larger than two millimeters is considered gravel. This category includes pebbles, cobbles and boulders. Like sandstone, when gravels are lithified they are considered conglomerates.

Conglomerates and Breccias

Conglomerate

Breccia. Notice the angular nature of the large clasts

Conglomerates are coarse grained rocks dominantly composed of gravel sized particles that are typically held together by a finer grained matrix. These rocks are often subdivided up into conglomerates and breccias. The major characteristic that divides these two categories is the amount of rounding. The gravel sized particles that make up conglomerates are well rounded while in breccias they are angular. Conglomerates are common in stratigraphic successions of most, if not all ages but only make up one percent or less, by weight of the total sedimentary rock mass. In terms or origin and depositional mechanisms they are very similar to sandstones. As a result, the two categories often contain the same sedimentary structures.

Sandstones

Sandstones are medium grained rocks composed of rounded or angular fragments of sand size, that often but not always have a cement uniting them together. These sand size particles are often quartz minerals but there are a few common categories and a wide variety of classification schemes that are given are used to classify sandstones based on composition. Classification schemes vary widely, but most geologists have adopted the Dott scheme, which uses the relative abundance of quartz, feldspar, and lithic framework grains and the abundance of muddy matrix between these larger grains.

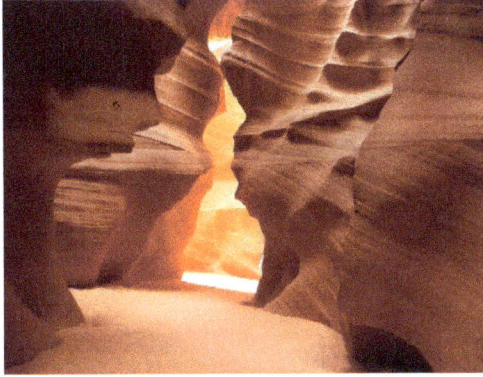

Sandstone from Lower Antelope Canyon

Mudrocks

Rocks that are classified as mudrocks are very fine grained. Silt and clay represent at least 50% of the material that mudrocks are composed of. Classification schemes for mudrocks tend to vary but most are based on the grain size of the major constituents. In mudrocks, these are generally silt, and clay.

According to Blatt, Middleton and Murray mudrocks that are composed mainly of silt particles are classified as siltstones. In turn, rocks that possess clay as the majority particle are called claystones. In geology, a mixture of both silt and clay is called mud. Rocks that possess large amounts of both clay and silt are called mudstones. In some cases the term shale is also used to refer to mudrocks and is still widely accepted by most. However, others have used the term shale to further divide mudrocks based on the percentage of clay constituents. The plate-like shape of clay allows its particles to stack up one on top of another creating laminae or beds. The more clay present in a given specimen, the more laminated a rock is. Shale, in this case, is reserved for mudrocks that are laminated, while mudstone refers those that are not.

Red mudrock

Black Shale

Diagenesis of Siliciclastic Sedimentary Rocks

Siliciclastic rocks initially form as loosely packed sediment deposits including gravels, sands, and muds. The process of turning loose sediment into hard is called lithification. During the process of lithification, sediments undergo physical, chemical and mineralogical changes before becoming rock. The primary physical process in lithification is compaction. As sediment transport and deposition continues, new sediments are deposited over top of previously deposited beds burying them. Burial continues and the weight of overlying sediments cause an increase in temperature and pressure. This increase in temperature and pressure causes loose grained sediments become tightly packed reducing porosity, essentially squeezing water out of the sediment. Porosity is further reduced by the precipitation of minerals into the remaining pore spaces. The final stage in the process is diagenesis and will be discussed in detail below.

Cementation

Cementation is the diagenetic process by which coarse clastic sediments become lithified or consolidated into hard, compact rocks, usually through the deposition or precipitation of minerals in the spaces between the individual grains of sediment. Cementation can occur simultaneously with deposition or at another time. Furthermore, once a sediment is deposited, it becomes subject to cementation through the various stages of diagenesis discussed below.

Shallow Burial (Eogenesis)

Eogenesis refers to the early stages of diagenesis. This can take place at very shallow depths, ranging from a few meters to tens of meters below the surface. The changes that occur during this diagenetic phase mainly relate to the reworking of the sediments. Compaction and grain repacking, bioturbation, as well as mineralogical changes all occur at varying degrees. Due to the shallow depths, sediments undergo only minor compaction and grain rearrangement during this stage. Organisms rework sediment near the depositional interface by burrowing, crawling, and in some cases sediment ingestion. This process can destroy sedimentary structures that were present upon deposition of the sediment. Structures such as lamination will give way to new structures associated the activity of organisms. Despite being close to the surface, eogenesis does provide conditions for important mineralogical changes to occur. This mainly involves the precipitation of new minerals.

Mineralogical Changes During Eogenesis

Mineralogical changes that occur during eogenesis as dependent on the environment in which that sediment has been deposited. For example, the formation of pyrite is characteristic of reducing conditions in marine environments. Pyrite can form as cement, or replace organic materials, such as wood fragments. Other important reactions include the formation of chlorite, glauconite, illite and iron oxide (if oxygenated pore water is present). The precipitation of potassium feldspar, quartz overgrowths, and carbonate cements also occurs under marine conditions. In non marine environments oxidizing conditions are almost always prevalent, meaning iron oxides are commonly produced along with kaolin group clay minerals. The precipitation of quartz and calcite cements may also occur in non marine conditions.

Deep Burial (Mesogenesis)

Compaction

As sediments are buried deeper, load pressures become greater resulting in tight grain packing and bed thinning. This causes increased pressure between grains thus increasing the solubility of grains. As a result, the partial dissolution of silicate grains occurs. This is called pressure solutions. Chemically speaking, increases in temperature can also cause chemical reaction rates to increase. This increases the solubility of most common minerals (aside from evaporites). Furthermore, beds thin and porosity decreases allowing cementation to occur by the precipitation of silica or carbonate cements into remaining pore space.

In this process minerals crystallize from watery solutions that percolate through the pores between grain of sediment. The cement that is produced may or may not have the same chemical composition as the sediment. In sandstones, framework grains are often cemented by silica or carbonate. The extent of cementation is dependent on the composition of the sediment. For example, in lithic sandstones, cementation is less extensive because pore space between framework grains is filled with a muddy matrix that leaves little space for precipitation to occur. This is often the case for mudrocks as well. As a result of compaction, the clayey sediments comprising mudrocks are relatively impermeable.

Dissolution

Dissolution of framework silicate grains and previously formed carbonate cement may occur during deep burial. Conditions that encourage this are essentially opposite of those required for cementation. Rock fragments and silicate minerals of low stability, such as plagioclase feldspar, pyroxenes, and amphiboles, may dissolve as a result of increasing burial temperatures and the presence of organic acids in pore waters. The dissolution of frame work grains and cements increases porosity particularly in sandstones.

Mineral Replacement

This refers to the process whereby one mineral is dissolved and a new mineral fills the space via precipitation. Replacement can be partial or complete. Complete replacement destroys the identity of the original minerals or rock fragments giving a biased view of the original mineralogy of the rock/ Porosity can also be affected by this process. For example, clay minerals tend to fill up pore space and thereby reducing porosity.

Telogenesis

In the process of burial, it is possible that siliciclastic deposits may subsequently be up-lifted as a result of a mountain building event or erosion. When uplift occurs, it exposes buried deposits to a radically new environment. Because the process brings material to or closer to the surface, sediments that undergo uplift are subjected to lower temperatures and pressures as well as slightly acidic rain water. Under these conditions, framework grains and cement are again subjected to dissolution and in turn increasing porosity. On the other hand, telogenesis can also change framework grains to clays, thus reducing porosity. These changes are dependent on the specific conditions that the rock is exposed as well as the composition of the rock and pore waters. Specific pore waters, can cause the further precipitation of carbonate or silica cements. This process can also encourage the process of oxidation on a variety of iron bearing minerals.

Sedimentary Breccias

Sedimentary breccias are a type of clastic sedimentary rock which are composed of angular to subangular, randomly oriented clasts of other sedimentary rocks. They may form either

1. in submarine debris flows, avalanches, mud flow or mass flow in an aqueous medium. Technically, turbidites are a form of debris flow deposit and are a fine-grained peripheral deposit to a sedimentary breccia flow.

2. as angular, poorly sorted, very immature fragments of rocks in a finer grained groundmass which are produced by mass wasting. These are, in essence, lith-ified colluvium. Thick sequences of sedimentary (colluvial) breccias are gener-ally formed next to fault scarps in grabens.

In the field, it may at times be difficult to distinguish between a debris flow sedimentary breccia and a colluvial breccia, especially if one is working entirely from drilling information. Sedimentary breccias are an integral host rock for many sedimentary exhalative deposits.

Igneous Clastic Rocks

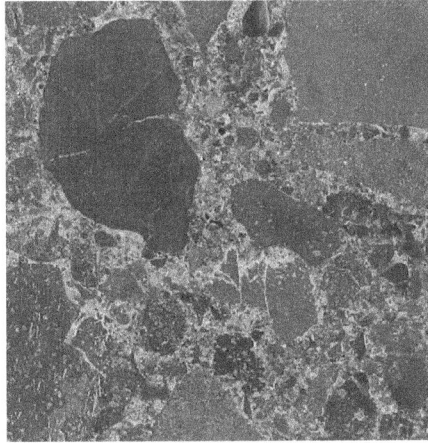

Basalt breccia, green groundmass is composed of epidote

Clastic igneous rocks include pyroclastic volcanic rocks such as tuff, agglomerate and intrusive breccias, as well as some marginal eutaxitic and taxitic intrusive morphologies. Igneous clastic rocks are broken by flow, injection or explosive disruption of solid or semi-solid igneous rocks or lavas.

Igneous clastic rocks can be divided into two classes:

1. Broken, fragmental rocks produced by intrusive processes, usually associated with plutons or porphyry stocks

2. Broken, fragmental rocks associated with volcanic eruptions, both of lava and pyroclastic type

Metamorphic Clastic Rocks

Clastic metamorphic rocks include breccias formed in faults, as well as some proto-mylonite and pseudotachylite. Occasionally, metamorphic rocks can be brecciated via hydrothermal fluids, forming a hydrofracture breccia.

Hydrothermal Clastic Rocks

Hydrothermal clastic rocks are generally restricted to those formed by hydrofracture, the process by which hydrothermal circulation cracks and brecciates the wall rocks and fills it in with veins. This is particularly prominent in epithermal ore deposits and is associated with alteration zones around many intrusive rocks, especially granites. Many skarn and greisen deposits are associated with hydrothermal breccias.

Impact Breccias

A fairly rare form of clastic rock may form during meteorite impact. This is composed primarily of ejecta; clasts of country rock, melted rock fragments, tektites (glass ejected from the impact crater) and exotic fragments, including fragments derived from the impactor itself.

Identifying a clastic rock as an impact breccia requires recognising shatter cones, tektites, spherulites, and the morphology of an impact crater, as well as potentially recognizing particular chemical and trace element signatures, especially osmiridium.

Mudrock

Mudrocks are a class of fine grained siliciclastic sedimentary rocks. The varying types of mudrocks include: *siltstone, claystone, mudstone, slate,* and *shale.* Most of the particles of which the stone is composed are less than 0.0625 mm (1/16th mm or 0.0025 inches) and are too small to study readily in the field. At first sight the rock types look quite similar; however, there are important differences in composition and nomenclature. There has been a great deal of disagreement involving the classification of mudrocks. There are a few important hurdles to classification, including:

1. Mudrocks are the least understood, and one of the most understudied sedimentary rocks to date

2. It is difficult to study mudrock constituents, due to their diminutive size and susceptibility to weathering on outcrops

3. And most importantly, there is more than one classification scheme accepted by scientists

Mudrocks make up fifty percent of the sedimentary rocks in the geologic record, and are easily the most widespread deposits on Earth. Fine sediment is the most abundant product of erosion, and these sediments contribute to the overall omnipresence of mudrocks. With increased pressure over time the platey clay minerals may become aligned, with the appearance of parallel layering (fissility). This finely bedded material that splits readily into thin layers is called *shale,* as distinct from *mudstone.* The lack of fissility or layering in mudstone may be due either to the original texture or to the disruption of layering by burrowing organisms in the sediment prior to lithification.

From the beginning of civilization, when pottery and mudbricks were made by hand, to now, mudrocks have been important. The first book on mudrocks, *Geologie des Argils* by Millot, was not published until 1964; however, scientists, engineers, and oil producers have understood the significance of mudrocks since the discovery of the Burgess Shale and the relatedness of mudrocks and oil. Literature on this omnipresent rock-type has been increasing in recent years, and technology continues to allow for better analysis.

Nomenclature

Mudrocks, by definition, consist of at least fifty percent mud-sized particles. Specifically, mud is composed of silt-sized particles that are between $1/16 - 1/256$ $((1/16)^2)$ of a millimeter in diameter, and clay-sized particles which are less than $1/256$ millimeter.

Mudrocks contain mostly clay minerals, and quartz and feldspars. They can also contain the following particles at less than 63 micrometres: calcite, dolomite, siderite, pyrite, marcasite, heavy minerals, and even organic carbon.

There are various synonyms for fine-grained siliciclastic rocks containing fifty percent or more of its constituents less than $1/256$ of a millimeter. Mudstones, shales, lutites, and argillites are common qualifiers, or umbrella-terms; however, the term mudrock has increasingly become the terminology of choice by sedimentary geologists and authors.

The term "mudrock" allows for further subdivisions of siltstone, claystone, mudstone, and shale. For example, a siltstone would be made of more than 50-percent grains that equate to $1/16 - 1/256$ of a millimeter. "Shale" denotes fissility, which implies an ability to part easily or break parallel to stratification. Siltstone, mudstone, and claystone implies lithified, or hardened, detritus without fissility.

Overall, "mudrocks" may be the most useful qualifying term, because it allows for rocks to be divided by its greatest portion of contributing grains and their respective grain size, whether silt, clay, or mud.

Type	Min grain	Max grain
Claystone	0 μm	4 μm
Mudstone	0 μm	64 μm
Siltstone	4 μm	64 μm
Shale	0 μm	64 μm
Slate	na	na

Claystone

Claystone in Slovakia

A claystone is lithified, and non-fissile mudrock. In order for a rock to be considered a claystone, it must consist of up to fifty percent clay, which measures less than 1/256 of a millimeter in particle size. Clay minerals are integral to mudrocks, and represent the first or second most abundant constituent by volume. There are 35 recognized clay mineral species on Earth. They make muds cohesive and plastic, or able to flow. Clay is by far the smallest particles recognized in mudrocks. Most materials in nature are clay minerals, but quartz, feldspar, iron oxides, and carbonates can weather to sizes of a typical clay mineral.

For a size comparison, a clay-sized particle is 1/1000 the size of a sand grain. This means a clay particle will travel 1000 times further at constant water velocity, thus requiring quieter conditions for settlement.

The formation of clay is well understood, and can come from soil, volcanic ash, and glaciation. Ancient mudrocks are another source, because they weather and disintegrate easily. Feldspar, amphiboles, pyroxenes, and volcanic glass are the principle donors of clay minerals.

Mudstone

A mudstone is a siliciclastic sedimentary rock that contains a mixture of silt- and clay-sized particles (at least 1/3 of each).

The terminology of "mudstone" is the Dunham classification scheme for limestones. In Dunham's classification, a mudstone is any limestone containing less than ten percent carbonate grains. Note, a siliciclastic mudstone does not deal with carbonate grains.

Friedman, Sanders, and Kopaska-Merkel (1992) suggest the use of "lime mudstone" to avoid confusion with siliciclastic rocks.

Siltstone

Siltstone at UAT, Estonia

A siltstone is a lithified, non-fissile mudrock. In order for a rock to be named a siltstone, it must contain over fifty percent silt-sized material. Silt is any particle smaller than sand, 1/16 of a millimeter, and larger than clay, 1/256 of millimeter. Silt is believed to be the product of physical weathering, which can involve freezing and thawing, thermal expansion, and release of pressure. Physical weathering does not involve any chemical changes in the rock, and it may be best summarised as the physical breaking apart of a rock.

One of the highest proportions of silt found on Earth is in the Himalayas, where phyllites are exposed to rainfall of up to five to ten meters (16 to 33 feet) a year. Quartz and feldspar are the biggest contributors to the silt realm, and silt tends to be non-cohesive, non-plastic, but can liquefy easily.

There is a simple test that can be done in the field to determine whether a rock is a siltstone or not, and that is to put the rock to one's teeth. If the rock feels "gritty" against one's teeth, then it is a siltstone.

Shale

Black Shale with pyrite

Shale is a fine grained, hard, laminated mudrock, consisting of clay minerals, and quartz and feldspar silt. Shale is lithified and fissile. It must have at least 50-percent of its particles measure less than 0.062 mm. This term is confined to argillaceous, or clay-bearing, rock.

There are many varieties of shale, including calcareous and organic-rich; however, black shale, or organic-rich shale, deserves further evaluation. In order for a shale to be a black shale, it must contain more than one percent organic carbon. A good source rock for hydrocarbons can contain up to twenty percent organic carbon. Generally, black shale receives its influx of carbon from algae, which decays and forms an ooze known as sapropel. When this ooze is cooked at desired pressure, three to six kilometers (1.8 - 3.7 miles) depth, and temperature, 90–120 °C (194–248 °F), it will form kerogen. Kerogen can be heated, and yield up to 10–150 US gallons (0.038–0.568 m³) of product per ton of rock.

Slate

Slate Roof

Slate is a hard mudstone that has undergone metamorphosis, and has well-developed cleavage. It has gone through metamorphism at temperatures between 200–250 °C (392–482 °F), or extreme deformation. Since slate is formed in the lower realm of metamorphism, based on pressure and temperature, slate retains its stratification and can be defined as a hard, fine-grained rock.

Slate is often used for roofing, flooring, or old-fashioned stone walls. It has an attractive appearance, and its ideal cleavage and smooth texture are desirable.

Creation of Mud and Mudrocks

Most mudrocks form in oceans or lakes, because these environments provide the quiet waters necessary for deposition. Although mudrocks can be found in every depositional environment on Earth, the majority are found in lakes and oceans.

Mud Transport and Supply

Heavy rainfall provides the kinetic motion necessary for mud, clay, and silt transport. Southeast Asia, including Bangladesh and India, receives high amounts of rain from monsoons, which then washes sediment from the Himalayas and surrounding areas to the Indian Ocean.

Warm, wet climates are best for weathering rocks, and there is more mud on ocean shelves off tropical coasts than on temperate or polar shelves. The Amazon system, for example, has the third largest sediment load on Earth, with rainfall providing clay, silt, and mud from the Andes in Peru, Ecuador, and Bolivia.

Rivers, waves, and longshore currents segregate mud, silt, and clay from sand and gravel due to fall velocity. Longer rivers, with low gradients and large watersheds, have the best carrying capacity for mud. The Mississippi River, a good example of long, low gradient river with a large amount of water, will carry mud from its northernmost sections, and deposit the material in its mud-dominated delta.

Mudrock Depositional Environments

Below is a listing of various environments that act as sources, modes of transportation to the oceans, and environments of deposition for mudrocks.

Alluvial Environments

Alluvial valleys, such as the Ganges in India, the Yellow in China, and Lower Mississippi in the United States are good examples of alluvial valleys. These systems have a continuous source of water, and can contribute mud through overbank sedimentation, when mud and silt is deposited overbank during flooding, and oxbow sedimentation where an abandoned stream is filled by mud.

In order for an alluvial valley to exist there must be a highly elevated zone, usually uplifted by active tectonic movement, and a lower zone, which acts as a conduit for water and sediment to the ocean.

Glaciers

Vast quantities of mud and till are generated by glaciations and deposited on land as till and in lakes. Glaciers can erode already susceptible mudrock formations, and this process enhances glacial production of clay and silt.

The Northern Hemisphere contains 90-percent of the world's lakes larger than 500 km (310 mi), and glaciers created many of those lakes. Lake deposits formed by glaciation, including deep glacial scouring, are abundant.

Non-glacial Lakes

Although glaciers formed 90-percent of lakes in the Northern Hemisphere, they are not responsible for the formation of ancient lakes. Ancient lakes are the largest and deepest in the world, and hold up to twenty percent of today's petroleum reservoirs. They are also the second most abundant source of mudrocks, behind marine mudrocks.

Ancient lakes owe their abundance of mudrocks to their long lives and thick deposits. These deposits were susceptible to changes in oxygen and rainfall, and offer a robust account of paleoclimate consistency.

Deltas

The Mississippi Delta

A delta is a subaerial or subaqueous deposit formed where rivers or streams deposit sediment into a water body. Deltas, such as the Mississippi and Congo, have massive potential for sediment deposit, and can move sediments into deep ocean waters. Delta environments are found at the mouth of a river, where its waters slow as they enter the ocean, and silt and clay are deposited.

Low energy deltas, which deposit a great deal of mud, are located in lakes, gulfs, seas, and small oceans, where coastal currents are also low. Sand and gravel-rich deltas are high-energy deltas, where waves dominate, and mud and silt are carried much farther from the mouth of the river.

Coastlines

Coastal currents, mud supply, and waves are a key factor in coastline mud deposition. The Amazon River supplies 500 million tons of sediment, which is mostly clay, to the coastal region of northeastern South America. 250 tons of this sediment moves along the coast and is deposited. Much of the mud accumulated here is more than 20 meters (65 feet) thick, and extends 30 kilometers (19 mi) into the ocean.

Much of the sediment carried by the Amazon can come from the Andes mountains, and the final distance traveled by the sediment is 6,000 km (3,700 mi).

Marine Environments

70-percent of the Earth's surface is covered by ocean, and marine environments are where we find the world's highest proportion of mudrocks. There is a great deal of lateral continuity found in the ocean, as opposed to continents which are confined.

In comparison, continents are temporary stewards of mud and silt, and the inevitable home of mudrock sediments is the oceans. Reference the mudrock cycle below in order to understand the burial and resurgence of the various particles

There are various environments in the oceans, including deep-sea trenches, abyssal plains, volcanic seamounts, convergent, divergent, and transform plate margins. Not only is land a major source of the ocean sediments, but organisms living within the ocean contribute, as well.

The world's rivers transport the largest volume of suspended and dissolved loads of clay and silt to the sea, where they are deposited on ocean shelves. At the poles, glaciers and floating ice drop deposits directly to the sea floor. Winds can provide fine grained material from arid regions, and explosive volcanic eruptions contribute as well. All of these sources vary in the rate of their contribution.

Sediment moves to the deeper parts of the oceans by gravity, and the processes in the ocean are comparable to those on land.

Location has a large impact on the types of mudrocks found in ocean environments. For example, the Apalachicola River, which drains in the subtropics of the United States, carries up to sixty to eighty percent kaolinite mud, whereas the Mississippi carries only ten to twenty percent kaolinite.

The Mudrock Cycle

We can imagine the beginning of a mudrock's life as sediment at the top of a mountain, which may have been uplifted by plate tectonics or propelled into the air from a volcano. This sediment is exposed to rain, wind, and gravity which batters and breaks apart the rock by weathering. The products of weathering, including particles ranging from clay to silt, to pebbles and boulders, are transported to the basin below, where it can solidify into one if its many sedimentary mudstone types.

Eventually, the mudrock will move its way kilometers below the subsurface, where pressure and temperature cook the mudstone into a metamorphosed gneiss. The metamorphosed gneiss will make its way to the surface once again as country rock or as magma in a volcano, and the whole process will begin again.

Important Properties

Color

Mudrocks form in various colors, including: red, purple, brown, yellow, green and grey, and even black. Shades of grey are most common in mudrocks, and darker colors of black come from organic carbons. Green mudrocks form in reducing conditions, where organic matter decomposes along with ferric iron. They can also be found in marine environments, where pelagic, or free-floating species, settle out of the water and decompose in the mudrock. Red mudrocks form when iron within the mudrock becomes oxidized, and depending on the intensity of red, one can determine if the rock has fully oxidized.

Fossils

Burgess Shale

Fossils are well preserved in mudrock formations, because the fine-grained rock protects the fossils from erosion, dissolution, and other processes of erosion. Fossils are particularly important for recording past environments. Paleontologists can look at a specific area and determine salinity, water depth, water temperature, water turbidity, and sedimentation rates with the aid of type and abundance of fossils in mudrock

One of the most famous mudrock formations is the Burgess Shale in Western Canada, which formed during the Cambrian. At this site, soft bodied creatures were preserved, some in whole, by the activity of mud in a sea. Solid skeletons are, generally, the only remnants of ancient life preserved; however, the Burgess Shale includes hard body parts such as bones, skeletons, teeth, and also soft body parts such as muscles, gills, and digestive systems. The Burgess Shale is one of the most significant fossil locations on Earth, preserving innumerable specimens of 500 million year old species, and its preservation is due to the protection of mudrock.

Another noteworthy formation is the Morrison Formation. This area covers 1.5 million square miles, stretching from Montana to New Mexico in the United States. It is considered one of the world's most significant dinosaur burial grounds, and its many fossils can be found in museums around the world. This site includes dinosaur fossils from a few dinosaur species, including the *Allosaurus, Diplodocus, Stegosaurus,* and *Brontosaurus*. There are also lungfish, freshwater mollusks, ferns and conifers. This deposit was formed by a humid, tropical climate with lakes, swamps, and rivers, which deposited mudrock. Inevitably, mudrock preserved countless specimens from the late Jurassic, roughly 150 million years ago.

Petroleum and Natural Gas

Mudrocks, especially black shale, are the source and containers of precious petroleum sources throughout the world. Since mudrocks and organic material require quiet water conditions for deposition, mudrocks are the most likely resource for petroleum. Mudrocks have low porosity, they are impermeable, and often, if the mudrock is not black shale, it remains useful as a seal to petroleum and natural gas reservoirs. In the case of petroleum found in a reservoir, the rock surrounding the petroleum is not the source rock, whereas black shale is a source rock.

Why Mudrocks are Important

As noted before, mudrocks make up fifty percent of the Earth's sedimentary geological record. They are widespread on Earth, and important for various industries.

Metamorphosed shale can hold emerald and gold, and mudrocks can host ore metals such as lead and zinc. Mudrocks are important in the preservation of petroleum and natural gas, due to their low porosity, and are commonly used by engineers to inhibit harmful fluid leakage from landfills.

Sandstones and carbonates record high-energy events in our history, and they are much easier to study. Interbedded between the high-energy events are mudrock formations that have recorded quieter, normal conditions in our Earth's history. It is the quieter, normal events of our geologic history we don't yet understand. Sandstones provide the big tectonic picture and some indications of water depth; mudrocks record oxygen content, a generally richer fossil abundance and diversity, and a much more informative geochemistry.

In recognition of mud and mudrocks' sometimes unappreciated importance to earth sciences, the Geological Society of London named 2015 as the "Year of Mud".

Phosphorite

Phosphorite, phosphate rock or rock phosphate is a non-detrital sedimentary rock which contains high amounts of phosphate bearing minerals. The phosphate content of phosphorite is at least 15 to 20%; if it is assumed that the phosphate minerals in phosphorite are hydroxyapatite and fluoroapatite, phosphate minerals contain roughly

18.5% phosphorus by weight and if phosphorite contains around 20% of these minerals, phosphorite is roughly 3.7% phosphorus by weight, which is a considerable enrichment over the typical sedimentary rock content of less than 0.2%. The phosphate is present as fluorapatite $Ca_5(PO_4)_3F$ typically in cryptocrystalline masses (grain sizes < 1 μm) referred to as collophane-sedimentary apatite deposits of uncertain origin. It is also present as hydroxyapatite $Ca_5(PO_4)_3OH$ or $Ca_{10}(PO_4)_6(OH)_2$, which is often dissolved from vertebrate bones and teeth, whereas fluorapatite can originate from hydrothermal veins. Other sources also include chemically dissolved phosphate minerals from igneous and metamorphic rocks. Phosphorite deposits often occur in extensive layers, which cumulatively cover tens of thousands of square kilometres of the Earth's crust.

Peloidal phosphorite, Phosphoria Formation, Simplot Mine, Idaho. 4.6 cm wide.

Fossiliferous peloidal phosphorite, (4.7 cm across), Yunnan Province, China.

Limestones and mudstones are common phosphate bearing rocks. Phosphate rich sedimentary rocks can occur in dark brown to black beds, ranging from centimeter sized laminae to beds that are several meters in thickness. Although these thick beds can exist they are rarely only composed of phosphatic sedimentary rocks. Phosphatic sedimentary rocks are commonly accompanied by or interbedded with shales, cherts, limestone, dolomites and sometimes sandstone. These layers contain the same textures and structures as fine grained limestones and may represent diagenetic replacements of carbonate minerals by phosphates. They also can be composed of peloids, ooids, fossils, and clasts that are made up of apatite. There are some phosphorites that are very small and have no distinctive granular textures. This means that their textures are similar to that of collophane, or fine micrite-like texture. Phosphatic grains may be accompanied by organic matter, clay minerals, silt sized detrital grains, and pyrite. Peloidal or pelletal phosphorites occur normally; whereas oolitic phosphorites are not common.

Phosphorites are known from Proterozoic banded iron formations in Australia, but are more common from Paleozoic and Cenozoic sediments. The Permian Phosphoria Formation of the western United States represents some 15 million years of sedimentation. It reaches a thickness of 420 metres and covers an area of 350,000 km². Commercially mined phosphorites occur in France, Belgium, Spain, Morocco, Tunisia and Algeria. In the United States phosphorites have been mined in Florida, Tennessee, Wyoming, Utah, Idaho and Kansas.

Classification of Phosphatic Sedimentary Rocks

(1) Pristine: Phosphates that are in pristine conditions have not undergone bioturbation. In other words, the word pristine is used when phosphatic sediment, phosphatized stromatolites and phosphate hardgrounds have not been disturbed.

(2) Condensed: Phosphatic particles, laminae and beds are considered condensed when they have been concentrated. This is helped by the extracting and reworking processes of phosphatic particles or bioturbation.

(3) Allochthonous: Phosphatic particles that were moved by turbulent or gravity-driven flows and deposited by these flows.

Phosphorus Cycle, Formation and Accumulation

The heaviest accumulation of phosphorus is mainly on the ocean floor. Phosphorus accumulation occurs from atmospheric precipitation, dust, glacial runoff, cosmic activity, underground hydrothermal volcanic activity, and deposition of organic material. However, the primary inflow of dissolved phosphorus is from continental weathering, brought out by rivers to the ocean. It is then processed by both micro- and macro-organisms. Diatomaceous plankton, phytoplankton, and zooplankton process and dissolve phosphorus in the water. The bones and teeth of certain fish (e.g. anchovies) absorb phosphorus and are later deposited and buried in the ocean sediment.

Depending on the pH and salinity levels of the ocean water, organic matter will decay and releases phosphorus from sediment in shallow basins. Bacteria and enzymes dissolve organic matter on the water bottom interface, thus returning phosphorus to the beginning of its biogenic cycle. Mineralization of organic matter can also cause the release of phosphorus back into the ocean water.

Depositional Environments

Phosphates are known to be deposited in a wide range of depositional environments. Normally phosphates are deposited in very shallow, near shore marine or low energy environments. This includes environments such as supratidal zones, littoral or intertidal zones, and most importantly estuarine. Currently, areas of oceanic upwelling cause the formation of phosphates. This is because of the constant stream of phosphate brought from the large, deep ocean reservoir. This cycle allows continuous growth of organisms.

Supratidal zones: Supratidal environments are part of the tidal flat system where the presence of strong wave activity is non-existent. Tidal flat systems are created along open coasts and relatively low wave energy environments. They can also develop on high energy coasts behind barrier islands where they are sheltered from the high energy wave action. Within the tidal flat system, the supratidal zone lies in a very high tide level. However, it can be flooded by extreme tides and cut across by tidal channels. This is also subaerially exposed, but is flooded twice a month by spring tides.

Littoral environments/ intertidal zones: Intertidal zones are also part of the tidal flat system. The intertidal zone is located within the mean high and low tide levels. It is subject to tidal shifts, which means that it is subaerially exposed once or twice a day. However, it is not exposed long enough to withhold vegetation. The zone contains both suspension sedimentation and bed load.

Estuarine environments: Estuarine environments, or estuaries, are located at the lower parts of rivers that flow into the open sea. Since they are in the seaward section of the drowned valley system they receive sediment from both marine and fluvial sources. These contain facies that are affected by tide and wave fluvial processes. An estuary is considered to stretch from the landward limit of tidal facies to the seaward limit of coastal facies. Phosphorites are often deposited in fjords within estuarine environments. These are estuaries with shallow sill constrictions. During Holocene sea-level rise, fjord estuaries formed by drowning of glacially-eroded U-shaped valleys.

The most common occurrence of phosphorites is related to strong marine upwelling of sediments. Upwelling is caused by deep water currents that are brought up to coastal surfaces where a large deposition of phosphorites may occur. This type of environment is the main reason why phosphorites are commonly associated with silica and chert. Estuaries are also known as a phosphorus "trap". This is because coastal estuaries contain a high productivity of phosphorus from marsh grass and benthic alge which allow an equilibrium exchange between living and dead organisms.

Types of Phosphorite Deposition

- Phosphate nodules: These are spherical concentrations that are randomly distributed along the floor of continental shelves. Most phosphorite grains are sand size although particles greater than 2 mm may be present. These larger grains, referred to as nodules, can range up to several tens of centimeters in size.

- Bioclastic phosphates or bone beds: Bone beds are bedded phosphate deposits that contain concentrations of small skeletal particles and coprolites. Some also contain invertebrate fossils like brachiopods and become more enriched in P_2O_5 after diagentic processes have occurred. Bioclastic phosphates can also be cemented by phosphate minerals.

- Phosphatization: Phosphatization is a type of rare diagenetic processes. It occurs when fluids that are rich in phosphate are leached from guano. These are then concentrated and reprecipitated in limestone. Phosphatized fossils or fragments of original phosphatic shells are important components within some these deposits.

Tectonic and Oceanographic Settings of Marine Phosphorites

- Epeiric sea phosphorites: Epeiric sea phosphorites are within marine shelf environments. These are in a broad and shallow cratonic setting. This is where granular phosphorites, phosphorite hardgrounds, and nodules occur.

- Continental margin phosphorites: Convergent, passive, upwelling, non-upwelling. This environment accumulates phosphorites in the form of hardgrounds, nodules and granular beds. These accumulate by carbonate fluorapatite percipitaion during early diagenesis in the upper few tens of centimeters of sediment. There are two different environmental conditions in which phosphorites are produced within continental margins. Continental margins can consist of organic rich sedimentation, strong coastal upwelling, and pronounced low oxygen zones. They can also form in conditions such as oxygen rich bottom waters and organic poor sediments.

- Seamount phosphorites: These are phosphorites that occur in seamounts, guyots, or flat topped seamounts, seamount ridges. These phosphorites are produced in association with iron and magnesium bearing crusts. In this setting the productivity of phosphorus is recycled within an iron oxidation reduction phosphorus cycle. This cycle can also form glauconite which is normally associated with modern and ancient phosphorites.

- Insular phosphorites: Insular phosphorites are located in carbonate islands, plateaus, coral island consisting of a reef surrounding a lagoon or, atoll lagoon, marine lakes. The phosphorite here originates from guano. Replacement of deep sea sediments precipitates, that has been formed in place on the ocean floor.

Production and Use

Guano phosphorite mining in the Chincha Islands of Peru, c. 1860

Phosphorite mine near Oron, Negev, Israel.

Deposits which contain phosphate in quantity and concentration which are economic to mine as ore for their phosphate content are not particularly common. The two main sources for phosphate are guano, formed from bird droppings, and rocks containing concentrations of the calcium phosphate mineral, apatite.

Phosphate rock is mined, beneficiated, and either solubilized to produce wet-process phosphoric acid, or smelted to produce elemental phosphorus. Phosphoric acid is reacted with phosphate rock to produce the fertilizer triple superphosphate or with anhydrous ammonia to produce the ammonium phosphate fertilizers. Elemental phosphorus is the base for furnace-grade phosphoric acid, phosphorus pentasulfide, phosphorus pentoxide, and phosphorus trichloride. Approximately 90% of phosphate rock production is used for fertilizer and animal feed supplements and the balance for industrial chemicals.

Froth flotation is used to concentrate the mined phosphorus to rock phosphate. The mined ore slurry is treated with fatty acids to cause calcium phosphate to become hydrophobic.

For general use in the fertilizer industry, phosphate rock or its concentrates preferably have levels of 30% phosphorus pentoxide (P_2O_5), reasonable amounts of calcium carbonate (5%), and <4% combined iron and aluminium oxides. Worldwide, the resources of high-grade ore are declining, and the beneficiation of lower grade ores by washing, flotation and calcining is becoming more widespread.

In addition to phosphate fertilisers for agriculture, phosphorus from rock phosphate is also used in animal feed supplements, food preservatives, anti-corrosion agents, cosmetics, fungicides, ceramics, water treatment and metallurgy.

As of 2006, the US is the world's leading producer and exporter of phosphate fertilizers, accounting for about 37% of world P_2O_5 exports. As of 2008, the world's total economic demonstrated resource of rock phosphate is 18 gigatonnes, which occurs principally as sedimentary marine phosphorites.

As of 2012, China, the United States and Morocco are the world's largest miners of phosphate rock, with a production of 77 megatonnes, 29.4 Mt and 26.8 Mt (including 2.5 Mt in Western Sahara) respectively in 2012 while Global production reached 195 Mt. Other countries with significant production include Brazil, Russia, Jordan and Tunisia. Historically, large amounts of phosphates were obtained from deposits on small islands such as Christmas Island and Nauru, but these sources are now largely depleted.

Iron-rich Sedimentary Rocks

Iron ore from Krivoy Rog, Ukraine

Iron-rich sedimentary rocks are sedimentary rocks which contain 15% or more iron. However, most sedimentary rocks contain iron in varying degrees. The majority of these rocks were deposited during specific geologic time periods: The Precambrian (3800 to 570 million years ago), the early Paleozoic (570 to 410 million years ago), and the middle to late Mesozoic (205 to 66 million years ago). Overall, they make up a very small portion of the total sedimentary record.

Iron-rich sedimentary rocks have economic uses as iron ores. Iron deposits have been located on all major continents with the exception of Antarctica. They are a major source of iron and are mined for commercial use. The main iron ores are from the oxide group consisting of hematite, goethite, and magnetite. The carbonate siderite is also typically mined. A productive belt of iron formations is known as an *iron range*.

Classification

The accepted classification scheme for iron-rich sedimentary rocks is to divide them into two sections: *ironstones* and *iron formations*

Ironstones

Ironstones consist of 15% iron or more in composition. This is necessary for the rock to even be considered an *iron-rich* sedimentary rock. Generally, they are from the Phanerozoic which means that they range in age from the present to 540 million years ago. They can contain iron minerals from the following groups: oxides, carbonates, and silicates. Some examples of minerals in iron-rich rocks containing oxides are limonite, hematite, and magnetite. An example of a mineral in iron-rich rock containing carbonates is siderite and an example of minerals in an iron-rich rock containing silicate is chamosite. They are often interbedded with limestones, shales, and fine-grained sandstones. They are typically nonbanded, however they can be very coarsely banded on occasion. They are hard and noncherty. The components of the rock range in size from sand to mud, but do not contain a lot of silica. They are also more aluminous. They are not laminated and sometimes contain *ooids*. Ooids can be a distinct characteristic though they are not normally a main component of ironstones. Within ironstones, ooids are made up of iron silicates and/or iron oxides and sometimes occur in alternating laminae. They normally contain fossil debris and sometimes the fossils are partly or entirely replaced by iron minerals. A good example of this is pyritization. They are smaller in size and less likely to be deformed or metamorphosed than iron formations. The term *iron ball* is occasionally used to describe an ironstone nodule.

Iron Formations

Iron formations must be at least 15% iron in composition, just like ironstones and all iron-rich sedimentary rocks. However, iron formations are mainly Precambrian in age which means that they are 4600 to 590 million years old. They are much older than ironstones. They tend to be cherty, though chert can not be used as a way to classify iron formations because it is a common component in many types of rocks. They are well banded and the banding can be anywhere from a few millimeters to tens of meters thick. The layers have very distinct banded successions that are made up of iron rich layers that alternate with layers of chert. Iron formations are often associates with dolomite, quartz-rich sandstone, and black shale. They sometimes grade locally into chert or dolomite. They can have a many different textures that resemble limestone. Some of these textures are micritic, pelleted, intraclastic, peloidal, oolitic, pisolitic, and stromatolitic. In low-grade iron formations, there are different dominant minerals dependent on the different types of facies. The dominant minerals in the oxide facies are magnetite and hematite. The dominant minerals in the silicate facies are greenalite, minnesotaite, and glauconite. The dominant mineral in the carbonate facies is siderite. The dominant mineral in the sulfide facies is pyrite. Most iron formations are deformed or metamorphosed simply due to their incredibly old age, but they still retain their unique distinctive chemical composition; even at high metamorphic grades. The higher the grade, the more metamorphosed it is. Low grade rocks may only be compacted while high grade rocks often can not be identified. They often contain a mixture of banded iron formations and granular iron formations. Iron formations can be divided into subdivisions known as: banded iron formations (BIFs) and granular iron formations (GIFs).

Bog ore

The above classification scheme is the most commonly used and accepted, though sometimes an older system is used which divides iron-rich sedimentary rocks into three categories: *bog iron deposits*, *ironstones*, and *iron formations*. A bog-iron deposit is iron that formed in a bog or swamp through the process of oxidation.

Banded Iron Formations Vs. Granular Iron Formations

Banded iron formation, North America

Banded iron formation close up, Upper Michigan

Banded Iron Formations

Banded iron formations (BIFs) were originally chemical muds and contain well developed thin lamination. They are able to have this lamination due to the lack of burrowers in the Precambrian. BIFs show regular alternating layers that are rich in iron and chert that range in thickness from a few millimeters to a few centimeters. The formation can continue uninterrupted for tens to hundreds of meters stratigraphically. These formations can contain sedimentary structures like cross-bedding, graded bedding, load casts, ripple marks, mud cracks, and erosion channels. In comparison to GIFs, BIFs contain a much larger spectrum of iron minerals, have more reduced facies, and are more abundant.

Granular Iron Formations

Granular iron formations (GIFs) were originally well-sorted chemical sands. They lack even, continuous bedding that takes the form of discontinuous layers. Discontinuous layers likely represent bedforms that were generated by storm waves and currents. Any layers that are thicker than a few meters and are uninterrupted, are rare for GIFs. They contain sand-sized clasts and a finer grained matrix, and generally belong to the oxide or silicate mineral facies.

Iron formations are sometimes divided into *Raptian-type*, *Algoma-type* and *Superior-type*.

Algoma Type

Algoma types are small lenticular iron deposits that are associated with volcanic rocks and turbidites. Iron content in this class type rarely exceeds 10^{10} tons. They range in thickness from 10-100 meters. Deposition occurs in island arc/back arc basins and intracratonic rift zones.

Superior Type

Superior types are large, thick, extensive iron deposits across stable shelves and in broad basins. Total iron content in this class type exceeds 10^{13} tons. They can extend to over 10^5 kilometers^2. Deposition occurs in relatively shallow marine conditions under transgressing seas.

Depositional Environment

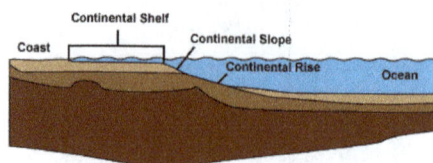

Profile illustrating the shelf, slope and rise

There are four facies types associated with iron-rich sedimentary rocks: oxide-, silicate-, carbonate-, and sulfide-facies. These facies correspond to water depth in a marine environment. Oxide-facies are precipitated under the most oxidizing conditions. Silicate- and carbonate-facies are precipitated under intermediate redox conditions. Sulfide-facies are precipitated under the most reducing conditions. There is a lack of iron-rich sedimentary rocks in shallow waters which leads to the conclusion that the depositional environment ranges from the continental shelf and upper continental slope to the abyssal plain. (The diagram does not have the abyssal plain labeled, but this would be located to the far right of the diagram at the bottom of the ocean).

Water colored by oxidized iron, Rio Tinto, Spain

Iron bacteria growing on iron-rich water seeping from a tall bluff, Sipsey Wilderness Area, Bankhead National Forest, AL

Chemical Reactions

Ferrous and ferric iron are components in many minerals, especially within sandstones. Fe^{2+} is in clay, carbonates, sulfides, and is even within feldspars in small amounts. Fe^{3+} is in oxides, hydrous, anhydrous, and in glauconites. Commonly, the presence of iron is determined to be within a rock due to certain colorations from oxidation. Oxidation is the loss of electrons from an element. Oxidation can occur from bacteria or by chemical oxidation. This often happens when ferrous ions come into contact with water (due to dissolved oxygen within surface waters) and a water-mineral reaction occurs. The formula for the oxidation/reduction of iron is:

$$Fe^{2+} \leftrightarrow Fe^{3+} + e^-$$

The formula works for oxidation to the right or reduction to the left.

Fe^{2+} is the ferrous form of iron. This form of iron gives up electrons easily and is a mild reducing agent. These compounds are more soluble because they are more mobile. Fe^{3+} is the ferric form of iron. This form of iron is very stable structurally because it's valence electron shell is half filled.

Laterization

Laterization is a soil forming process that occurs in warm and moist climates under broadleaf evergreen forests. Soils formed by laterization tend to be highly weathered with high iron and aluminium oxide content. Goethite is often made from this process and is a major source of iron in sediments. However, once it is deposited it must be dehydrated in order to come to an equilibrium with hematite. The dehydration reaction is:

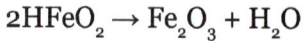

$$2HFeO_2 \rightarrow Fe_2O_3 + H_2O$$

Pyritized *Lytoceras*

Ankerite, Mineralogical Museum, Bonn, Germany

Pyritization

Pyritization is discriminatory. It rarely happens to soft tissue organisms and aragonitic fossils are more susceptible to it than calcite fossils. It commonly takes place in marine depositional environments where there is organic material. The process is caused by sulfate reduction which replaces carbonate skeletons (or shells) with pyrite (FeS_2). It generally does not preserve detail and the pyrite forms within the structure as many microcrystals. In fresh water environments, siderite will replace carbonate shells instead of pyrite due to the low amounts of sulfate. The amount of pyritization that has taken place within a fossil may sometimes be referred to as degree of pyritization (DOP).

Iron Minerals

- *Ankerite* ($Ca(Mg,Fe)(CO_3)_2$) and siderite ($FeCO_3$) are carbonates and favor alkaline, reducing conditions. They commonly occur as concretions in mudstones and siltstones.

- *Pyrite* and *marcasite* (FeS_2) are sulfide minerals and favor reducing conditions. They are the most common in fine-grained, dark colored mudstones.

- *Hematite* (Fe_2O_3) is usually the pigment in red beds and requires oxidizing conditions.

- *Limonite* ($2Fe_2O_3 \cdot 3H_2O$) is used for unidentified massive hydroxides and oxides of iron.

Bournonite with a pyrite crystal matrix, Chichibu Mine, Nakatsugawa, Honshu Island, Japan

Oolitic Hematite, Clinton, Oneida County, NY

Limonite, USGS

Iron–rich Rocks in Thin Section

Thin section of rhyolite volcanic rock showing an oxidized iron matrix (orange/brown color)

Magnetite and hematite are opaque under the microscope under transmitted light. Under reflected light, magnetite shows up as metallic and a silver or black color. Hematite will be a more reddish-yellow color. Pyrite is seen as opaque, a yellow-gold color, and metallic. Chamosite is an olive-green color in thin section that readily oxidizes to limonite. When it is partially or fully oxidized to limonite, the green color becomes a yellow-ish brown. Limonite is opaque under the microscope as well. Chamosite is an iron silicate and it has a birefringence of almost zero. Siderite is an iron carbonate and it has a very high birefringence. The thin sections often reveal marine fauna within oolitic ironstones. In older samples, the ooids may be squished and have hooked tails on either end due to compaction.

Carbonate Rock

Carbonate rocks are a class of sedimentary rocks composed primarily of carbonate minerals. The two major types are limestone, which is composed of calcite or aragonite (different crystal forms of $CaCO_3$) and dolostone, which is composed of the mineral dolomite ($CaMg(CO_3)_2$).

Calcite can be either dissolved by groundwater or precipitated by groundwater, depending on several factors including the water temperature, pH, and dissolved ion concentrations. Calcite exhibits an unusual characteristic called retrograde solubility in which it becomes less soluble in water as the temperature increases.

Carbonate ooids on the surface of a limestone; Carmel Formation (Middle Jurassic) of southern Utah, USA. Largest is 1.0 mm in diameter.

When conditions are right for precipitation, calcite forms mineral coatings that cement the existing rock grains together or it can fill fractures.

Karst topography and caves develop in carbonate rocks because of their solubility in dilute acidic groundwater. Cooling groundwater or mixing of different groundwaters will also create conditions suitable for cave formation.

Marble is the metamorphic carbonate rock. Rare igneous carbonate rocks exist as intrusive carbonatites and even rarer volcanic carbonate lava.

Conglomerate (Geology)

Carmelo Formation (Conglomerate) at Point Lobos

Conglomerate is a coarse-grained clastic sedimentary rock that is composed of a substantial fraction of rounded to subangular gravel-size clasts, e.g., granules, pebbles, cobbles, and boulders, larger than 2 mm (0.079 in) in diameter. Conglomerates form by the consolidation and lithification of gravel. Conglomerates typically contain finer grained sediment, e.g., either sand, silt, clay or combination of them, called *matrix* by geologists, filling their interstices and are often cemented by calcium carbonate, iron oxide, silica, or hardened clay.

The size and composition of the gravel-size fraction of a conglomerate may or may not vary in composition, sorting, and size. In some conglomerates, the gravel-size class consist almost entirely of what were clay clasts at the time of deposition. Conglomerates can be found in sedimentary rock sequences of all ages but probably make up less than 1 percent by weight of all sedimentary rocks. In terms of origin and depositional mechanisms, they are closely related to sandstones and exhibit many of the same types of sedimentary structures, e.g., tabular and trough cross-bedding and graded bedding.

Classification

Conglomerates may be named and classified by the:

- Amount and type of matrix present

- Composition of gravel-size clasts they contain

- Size range of gravel-size clasts present

The classification method depends on the type and detail of research being conducted.

A sedimentary rock composed largely of gravel is first named according to the roundness of the gravel. If the gravel clasts that comprise it is largely well-rounded to subrounded, it is a *conglomerate*. If the gravel clasts that comprise it are largely angular, it is a breccia. Such breccias can be called sedimentary breccias to differentiate them from other types of breccia, e.g. volcanic and fault breccias. Sedimentary rocks that contain a mixture of rounded and angular gravel clasts are sometimes called breccio-conglomerate.

Texture

Conglomerates are rarely composed entirely of gravel-size clasts. Typically, the space between the gravel-size clasts is filled by a mixture composed of varying amounts of silt, sand, and clay, known as *matrix*. If the individual gravel clasts in a conglomerate are separated from each other by an abundance of matrix such that they are not in contact with each other and *float* within the matrix, it is called a paraconglomerate. Paraconglomerates are also often unstratified and can contain more matrix than gravel clasts. If the gravel clasts of a conglomerate are in contact with each other, it is called a orthoconglomerate. Unlike paraconglomerates, orthoconglomerates are typically cross-bed-ded and often well-cemented and lithified by either calcite, hematite, quartz, or clay.

The differences between paraconglomerates and orthoconglomerates reflect differenc-es in how they are deposited. Paraconglomerates are commonly either glacial tills or debris flow deposits. Orthoconglomerates are typically associated with aqueous cur-rents of some sort.

A conglomerate at the base of the Cambrian in the Black Hills, South Dakota.

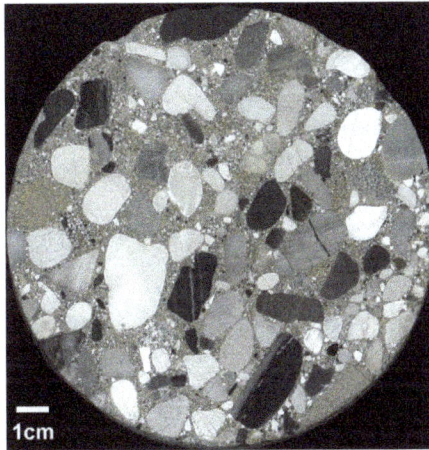

Section of polymict conglomerate from offshore rock core, Alaska, approximate depth 10,000 ft.

Conglomerate and sandstone alternating layers - Nerriga, New South Wales, Australia. Approx 5m vertical wall of road cut (vertical marks by construction machine.)

Clast Composition

Conglomerates are also classified according to the composition of their clasts. A conglomerate or any clastic sedimentary rock that consists of a single rock or mineral is known as either a monomict, monomictic, oligomict, or oligomictic conglomerate. If the conglomerate consists of two or more different types of rocks, minerals, or combination of both, it is known as either a polymict or polymictic conglomerate. If a polymictic conglomerate contains an assortment of the clasts of metastable and unstable rocks and minerals, it called either a petromict or petromictic conglomerate.

In addition, conglomerates are classified by source as indicated by the lithology of the gravel-size clasts If these clasts consist of rocks and minerals that are significantly different in lithology from the enclosing matrix and, thus, older and derived from outside the basin of deposition, the conglomerate is known as an extraformational conglomerate. If these clasts consist of rocks and minerals that are identical to or consistent with the lithology of the enclosing matrix and, thus, penecontemporaneous and derived from within the basin of deposition, the conglomerate is known as an intraformational conglomerate.

Two recognized types of type of intraformational conglomerates are shale-pebble and flat-pebble conglomerates. A shale-pebble conglomerate is a conglomerate that is composed largely of clasts of rounded mud chips and pebbles held together by clay minerals and created by erosion within environments such as within a river channel or along a lake margin. Flat-pebble conglomerates (edgewise conglomerates) are conglomerates that consist of relatively flat clasts of lime mud created by either storms or tsunami eroding a shallow sea bottom or tidal currents eroding tidal flats along a shoreline.

Clast Size

Finally, conglomerates are often differentiated and named according to the dominant clast size comprising them. In this classification, a conglomerate composed largely of granule-size clasts would be called a granule conglomerate; a conglomerate composed largely of pebble-size clasts would be called a pebble conglomerate; and a conglomerate composed largely of cobble-size clasts would be called a cobble conglomerate.

Sedimentary Environments

Conglomerates are deposited in a variety of sedimentary environments.

Deepwater Marine

In turbidites, the basal part of a bed is typically coarse-grained and sometimes conglomeratic. In this setting, conglomerates are normally very well sorted, well-rounded and often with a strong A-axis type imbrication of the clasts.

Shallow Marine

Conglomerates are normally present at the base of sequences laid down during marine transgressions above an unconformity, and are known as *basal conglomerates*. They represent the position of the shoreline at a particular time and are diachronous.

Fluvial

Conglomerates deposited in fluvial environments are typically well rounded and well sorted. Clasts of this size are carried as bedload and only at times of high flow-rate. The maximum clast size decreases as the clasts are transported further due to attrition, so conglomerates are more characteristic of immature river systems. In the sediments deposited by mature rivers, conglomerates are generally confined to the basal part of a channel fill where they are known as *pebble lags*. Conglomerates deposited in a fluvial environment often have an AB-plane type imbrication.

Alluvial

Fanglomerate

Alluvial deposits form in areas of high relief and are typically coarse-grained. At mountain fronts individual alluvial fans merge to form braidplains and these two environments are associated with the thickest deposits of conglomerates. The bulk of conglomerates deposited in this setting are clast-supported with a strong AB-plane imbrication. Matrix-supported conglomerates, as a result of debris-flow deposition, are quite commonly associated with many alluvial fans. When such conglomerates accumulate within an alluvial fan, in rapidly eroding (e.g., desert) environments, the resulting rock unit is often called a fanglomerate.

Glacial

Glaciers carry a lot of coarse-grained material and many glacial deposits are conglomeratic. Tillites, the sediments deposited directly by a glacier, are typically poorly sorted, matrix-supported conglomerates. The matrix is generally fine-grained, consisting of finely milled rock fragments. Waterlaid deposits associated with glaciers are often conglomeratic, forming structures such as eskers.

Examples

An example of conglomerate can be seen at Montserrat, near Barcelona. Here, erosion has created vertical channels that give the characteristic jagged shapes the mountain is named for (Montserrat literally means "jagged mountain"). The rock is strong enough to use as a building material, as in the Santa Maria de Montserrat Abbey.

Another example, the Crestone Conglomerate, occurs in and near the town of Crestone, at the foot of the Sangre de Cristo Range in Colorado's San Luis Valley. The Crestone Conglomerate consists of poorly sorted fanglomerates that accumulated in prehistoric alluvial fans and related fluvial systems. Some of these rocks have hues of red and green.

Conglomerate cliffs are found on the east coast of Scotland from Arbroath northwards along the coastlines of the former counties of Angus and Kincardineshire. Dunottar Castle sits on a rugged promontory of conglomerate jutting into the North Sea just south of the town of Stonehaven.

Conglomerate may also be seen in the domed hills of Kata Tjuta, in Australia's Northern Territory.

In the nineteenth century a thick layer of Pottsville conglomerate was recognized to underlie anthracite coal measures in Pennsylvania.

Examples on Mars

On Mars, slabs of conglomerate have been found at an outcrop named "Hottah", and have been interpreted by scientists as having formed in an ancient streambed. The gravels, which were discovered by NASA's Mars rover Curiosity, range from the size of sand particles to the size of golf balls. Analysis has shown that the pebbles were deposited by a stream that flowed at walking pace and was ankle- to hip-deep.

Metaconglomerate

Metamorphic alteration transforms conglomerate into metaconglomerate.

Sedimentary Structures

Sedimentary structures are those structures formed during sediment deposition.

Sedimentary structures such as cross bedding, graded bedding and ripple marks are utilized in stratigraphic studies to indicate original position of strata in geologically complex terrains and understand the depositional environment of the sediment.

Flow Structures

Megaripple/dune, formed in the upper flow regime, from Utah.

There are two kinds of flow structures: bidirectional (multiple directions, back-and-forth) and unidirectional. Flow regimes in single-direction (typically fluvial) flow, which at varying speeds and velocities produce different structures, are called bedforms. In the *lower flow regime*, the natural progression is from a flat bed, to some sediment movement (saltation etc.), to ripples, to slightly larger dunes. Dunes have a vortex in the lee side of the dune. As the *upper flow regime* forms, the dunes become flattened out, and then produce antidunes. At higher still velocity, the antidunes are flattened and most sedimentation stops, as erosion takes over as the dominant process.

Bedforms Vs. Flow

Typical unidirectional bedforms represent a specific flow velocity, assuming typical sediments (sands and silts) and water depths, and a chart such as below can be used for interpreting depositional environments, with increasing water velocity going down the chart.

Flow Regime	Bedform	Preservation Potential	Identification Tips
	Lower plane bed	High	Flat laminae, almost lack of current
Lower	Ripple marks	Relatively Low	Small, cm-scale undulations
	Sand waves	Medium to low	Rare, longer wavelength than ripples
	Dunes/Megaripples	Low	Large, meter-scale ripples
	Upper plane bed	High	Flat laminae, +/- aligned grains (parting lineations)
Upper	Antidunes	Low	Water in phase with bedform, low angle, subtle laminae
	Pool and chute	Very low	Mostly erosional features

Ripple Marks

Wave ripple or symmetric ripple, from Permian rocks in Nomgon, Mongolia. Note "decapitation" of ripple crests due to change in current.

Ripple marks usually form in conditions with flowing water, in the lower part of the Lower Flow Regime. There are two types of ripple marks:

- Symmetrical ripple marks - Often found on beaches, they are created by a two way current, for example the waves on a beach (swash and backwash). This creates ripple marks with pointed crests and rounded troughs, which aren't inclined more to a certain direction. Three common sedimentary structures that are created by these processes are herringbone cross-stratification, flaser bedding, and interference ripples.

- Asymmetrical ripple marks - These are created by a one way current, for example in a river, or the wind in a desert. This creates ripple marks with still pointed crests and rounded troughs, but which are inclined more strongly in the direction of the current. For this reason, they can be used as palaeocurrent indicators.

Antidunes

Antidunes are the sediment bedforms created by fast, shallow flows of water with a Froude number greater than 1. Antidunes form beneath standing waves of water that periodically steepen, migrate, and then break upstream. The antidune bedform is characterized by shallow foresets, which dip upstream at an angle of about ten degrees that can be up to five meters in length. They can be identified by their low angle foresets. For the most part, antidunes bedforms are destroyed during decreased flow, and therefore cross bedding formed by antidunes will not be preserved.

Biological Structures

A number of biologically-created sedimentary structures exist, called trace fossils. Examples include burrows and various expressions of bioturbation. Ichnofacies are groups of trace fossils that together help give information on the depositional environment. In general, as deeper (into the sediment) burrows become more common, the shallower the water. As (intricate) surface traces become more common, the water becomes deeper.

Microbes may also interact with sediment to form Microbially Induced Sedimentary Structures.

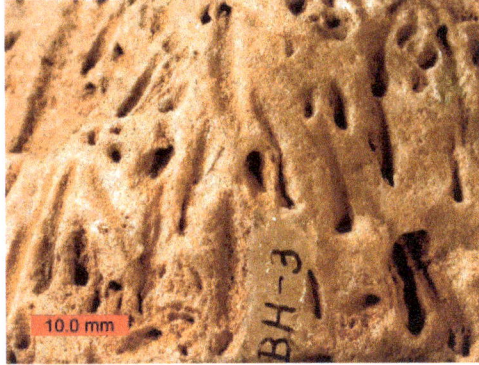

Skolithos trace fossil. Scale bar is 10 mm.

Soft Sediment Deformation Structures

Soft sediment deformation in Dead Sea sediments, Israel. Possibly a seismite.

Soft-sediment deformation structures or SSD, is a consequence of the loading of wet sediment as burial continues after deposition. The heavier sediment "squeezes" the water out of the underlying sediment due to its own weight. There are three common variants of SSD:

- load structures or load casts (also a type of sole marking) are blobs that form when a denser, wet sediment slumps down on and into a less dense sediment below.

- pseudonodules or ball-and-pillow structures, are pinched-off load structures; these may also be formed by earthquake energy and referred to as seismites.

- flame structures, "fingers" of mud that protrude into overlying sediments.

- clastic dikes are seams of sedimentary material that cut across sedimentary strata.

Bedding Plane Structures

Flute cast from Book Cliffs area, Utah.

Mudcracks in rock at Roundtop Hill, Maryland.

Bedding Plane Structures are commonly used as paleocurrent indicators. They are formed when sediment has been deposited and then reworked and reshaped. They include:

- Sole markings form when an object gouges the surface of a sedimentary layer; this groove is later preserved as a cast when filled in by the layer above. They include:

    - Flute casts are scours dug into soft, fine sediment which typically get filled by an overlying bed. Measuring the long axis of the flute cast gives the direction of flow, with the scoop-shaped end pointing in the upcurrent direction and the tapered end pointing downcurrent (paleoflow direction). The convexity of the flute cast also points stratigraphically down.

    - Tool marks are a type of sole marking formed by grooves left in a bed by objects dragged along by a current. The average direction of these can be assumed to be the axis of flow direction.

- Mudcracks form when mud is dewatered, shrinks, and leaves a crack. This tells you that the mud was saturated with water and then exposed to air. Mudcracks curl upwards, so they can be used as geopetal structures. Syneresis cracks form in a similar way, with the exception that they are never exposed to air, instead being caused by changes in the salinity of the surrounding water.

- Raindrop impressions form on exposed sediment by raindrop impacts.

- Parting lineations are subtly aligned minerals that form in the lower part of the Upper Flow Regime within plane beds.

Within Bedding Structures

A teepee structure in modern halite deposits along the western shore of the Dead Sea, Israel.

These structures are within sedimentary bedding and can help with the interpretation of depositional environment and paleocurrent directions. They are formed when the sediment is deposited.

- Cross bedding - This can include ripples and dunes, or any cross stratification caused by currents. The "cross" refers to the angle between flat bedding and the inclined bedding of the cross bed, typically about 34 degrees. Paleocurrents are best found from cross beds that have 3D architecture exposed so you can measure the axis of the trough of the cross bed.

- Hummocky cross-stratification is made up of undulating sets of cross-laminae that are concave-up (swales) and convex-up (hummocks). These cross-beds gently cut into each other with curved erosional surfaces. They form in shallow-water, storm-dominated environments. Strong storm-wave action erodes the seabed into low hummocks and swales that lack a specific orientation.

- Imbrication is the stacking of larger clasts in the direction of flow.

- Normal graded bedding occurs when current velocity changes and grains are progressively dropped out of the current. The most common place to find this is in a turbidite deposit. This can also be inverted, called reversed graded bedding, and is common in debris flows.

- Bioturbation - Biological stirring of sediment (i.e. burrowing); typical of shallow water, finer-grained sediment.

- Tidal bundle - variation in bedding thickness in a tidal environment caused by alternation of spring and neap tides

Secondary Sedimentary Structures

Secondary sedimentary structures form after primary deposition occurs or, in some cases, during the diagenesis of a sedimentary rock. Common secondary structures include any form of bioturbation, soft-sediment deformation, teepee structures, root-traces, and soil mottling. Liesegang rings, cone-in-cone structures, raindrop impressions, and vegetation-induced sedimentary structures would also be considered secondary structures.

Cross-bedding

Cross bedding forms during deposition on the inclined surfaces of bedforms such as ripples and dunes, and indicates that the depositional environment contained a flowing medium (typically water or wind). Examples of these bedforms are ripples, dunes, anti-dunes, sand waves, hummocks, bars, and delta slopes. Cross-bedding is widespread in many environments. Environments in which water movement is fast enough and deep enough to develop large-scale bed forms fall into three natural groupings: rivers, tide-dominated coastal and marine settings.

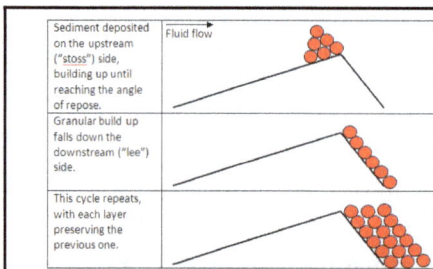

Sediment deposited on the upstream ("stoss") side, building up until reaching the angle of repose. / Granular build up falls down the downstream ("lee") side. / This cycle repeats, with each layer preserving the previous one.	
Formation of cross-stratification	Cross-bedding in a sandstone dome in the Canyons of the Escalante.
Tabular cross bedding in the South Bar Formation in Nova Scotia	Sand dune cross-beds can be large, such as in the Jurassic-age erg deposits of the Navajo Sandstone in Canyonlands National Park.

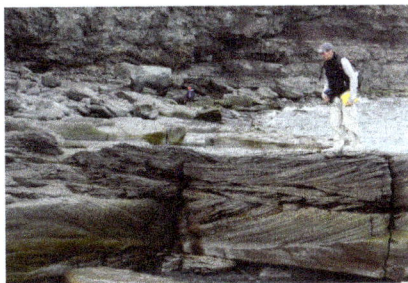

Trough cross bedding in the Waddens Cove Formation in Nova Scotia

Trough cross bedding in the Lower Cove Formation in Nova Scotia

In geology, the sedimentary structures known as cross-bedding are the (near-) horizontal units that are internally composed of inclined layers. This is a case in geology in which the original depositional layering is tilted, and the tilting is not a result of post-depositional deformation. Cross-beds or "sets" are the groups of inclined layers, and the inclined layers are known as cross strata.

Significance

Cross beds can tell geologists much about what an area was like in ancient times. The direction the beds are dipping indicates paleocurrent, the rough direction of sediment transport. The type and condition of sediments can tell geologists the type of environment (rounding, sorting, composition...). Studying modern analogs allows geologists to draw conclusions about ancient environments. Paleocurrent can be determined by seeing a cross-section of a set of cross-beds. However, to get a true reading, the axis of the beds must be visible. It is also difficult to distinguish between the cross beds of a dune and the cross beds of an antidune. This could lead to misinterpretation since dunes dip downstream while antidunes dip upstream.

The direction of motion of the cross-beds can show ancient flow or wind directions (called paleocurrents). The foresets are deposited at the angle of repose (~34 degrees from the horizontal), so geologists are able to measure dip direction of the cross-bedded sediments and calculate the paleoflow direction. However, most cross-beds are not tabular, they are troughs. Since troughs can give a 180 degree variation of the dip of foresets, false paleocurrents can be taken by blindly measuring foresets. In this case, true paleocurrent direction is determined by the axis of the trough. Paleocurrent direction is important in reconstructing past climate and drainage patterns: sand dunes preserve the prevalent wind directions, and current ripples show the direction rivers were moving.

Formation

Cross-bedding is formed by the downstream migration of bedforms such as ripples or dunes in a flowing fluid. The fluid flow causes sand grains to saltate up the upstream ("stoss") side of the bedform and collect at the peak until the angle of repose is reached. At this point, the crest of granular material has grown too large and will be overcome by the force of the depositing fluid, falling down the downstream ("lee") side of the dune. Repeated avalanches will eventually form the sedimentary structure known as cross-bedding, with the structure dipping in the direction of the paleocurrent.

The sediment that goes on to form cross-stratification is generally sorted before and during deposition on the "lee" side of the dune, allowing cross strata to be recognized in rocks and sediment deposits.

The angle and direction of cross-beds are generally fairly consistent. Individual cross-beds can range in thickness from just a few tens of centimeters, up to hundreds of feet or more depending upon the depositional environment and the size of the bedform. Cross-bedding can form in any environment in which a fluid flows over a bed with mobile material. It is most common in stream deposits (consisting of sand and gravel), tidal areas, and in aeolian dunes.

Internal Sorting Patterns

Cross-bedded sediments are recognized in the field by the many layers of "foresets", which are the series of layers that form on the lee side of the bedform (ripple or dune). These foresets are individually differentiable because of small-scale separation between layers of material of different sizes and densities.

Cross-bedding can also be recognized by truncations in sets of ripple foresets, where previously-existing stream deposits are eroded by a later flood, and new bedforms are deposited in the scoured area.

Geometries

Cross-bedding can be subdivided according to the geometry of the sets and cross strata into subcategories. The most commonly described types are tabular cross-bedding and trough cross-bedding. Tabular cross-bedding, or planar bedding consists of cross-bedded units that are extensive horizontally relative to the set thickness and that have essentially planar bounding surfaces. Trough cross-bedding, on the other hand, consists of cross-bedded units in which the bounding surfaces are curved, and hence limited in horizontal extent.

Tabular (Planar) Cross-beds

Tabular (planar) cross-beds consist of cross-bedded units that are large in horizontal extent relative to set thickness and that have essentially planar bounding surfaces. The foreset laminae of tabular cross-beds are curved so as to become tangential to the basal surface.

Tabular cross-bedding is formed mainly by migration of large-scale, straight-crested ripples and dunes. It forms during lower-flow regimes. Individual beds range in thickness from a few tens of centimeters to a meter or more, but bed thickness down to 10 centimeters has been observed. Where the set height is less than 6 centimeters and the cross-stratification layers are only a few millimeters thick, the term cross-lamination is used, rather than cross-bedding. Cross bed sets occur typically in granular sediments, especially sandstone, and indicate that sediments were deposited as ripples or dunes, which advanced due to a water or air current.

Trough Cross-beds

Cross beds are layers of sediment that are inclined relative to the base and top of the bed they are associated with. Cross beds can tell modern geologists many things about ancient environments such as- depositional environment, the direction of sediment transport (paleocurrent) and even environmental conditions at the time of deposition. Typically, units in the rock record are referred to as beds, while the constituent layers that make up the bed are referred to as laminae, when they are less than 1 cm thick and strata when they are greater than 1 cm in thickness. Cross beds are angled relative to either the base or the top of the surrounding beds. As opposed to angled beds, cross beds are deposited at an angle rather than deposited horizontally and deformed later on. Trough cross-beds have lower surfaces which are curved or scoop shaped and truncate the underlying beds. The foreset beds are also curved and merge tangentially with the lower surface. They are associated with sand dune migration.

Sediment

The shape of the grains and the sorting and composition of sediment can provide additional information on the history of cross beds. Roundness of the grains, limited variation in grain size, and high quartz contents are generally attributed to longer histories of weathering and sediment transport. For example: well-rounded, and well-sorted sand that is mostly composed of quartz grains is commonly found in beach environments, far from the source of the sediment. Poorly sorted and angular sediment that is composed of a diversity of minerals is more commonly found in rivers, near the source of the sediment. However, older sedimentary deposits are frequently eroded and re-mobilized. Thus, a river may well erode an older formation of well-rounded, well-sorted beach sands of nearly pure quartz.

Environments

Rivers

Flows are characterized by climate (snows, rain, and ice melting) and gradient. Discharge variations measured on a variety of time scales can change water depth, and speed. Some rivers can be characterized by a predictable seasonally controlled hydrograph (reflecting snow melt or rainy season). Others are dominated by durational variations characteristic of alpine glaciers run-off or random storm events, which produce flashy discharge. Few rivers have a long term record of steady flow in the rock record.

Bed forms are relatively dynamic sediment storage bodies with response times that are short relative to major changes in flow characteristics. Large scale bed forms are periodic and occur in the channel (scaled to depth). Their presence and morphologic variability have been related to flow strength expressed as mean velocity or shear stress.

In a fluvial environment, the water in a stream loses energy and its ability transport sediment. The sediment "falls" out of the water and is deposited along a point bar. Over time the river may dry up or avulse and the point bar may be preserved as cross bedding.

Tide-dominated

Tide dominated environments include:

- Coastal water bodies that are partially enclosed by topography, yet have a free connection to the sea.

- Coast lines that have a tidal range of greater than one meter.

- Areas in which the water run-off volume is low relative to the tidal volume or impact.

In general, the greater the tidal range the greater the maximum flow strength. Cross-stratification in tidal-dominated areas can lead to the formation of Herringbone cross-stratification.

Although the flow direction reverses regularly, the flow patterns of flood on ebb currents commonly do not coincide. Consequently, the water and transport sediment may follow a roundabout route in and out of the estuary. This leads to spatially varied systems where some parts of the estuary are flood dominated and other parts are ebb dominated. The temporal and spatial variability of flow and sediment transport, coupled with regular fluctuating water levels creates a variety of bed form morphology.

Shallow Marine

Large scale bed forms occur on shallow, terrigenous or carbonate clastic continental shelves and epicontinental platforms which are affected by strong geostaphic currents, occasional storm surges and/or tide currents.

Aeolian

In an aeolian environment, cross beds often exhibit inverse grading due to their deposition by grain flows. Winds blow sediment along the ground until they start to accumulate. The side that the accumulation occurs on is called the windward side. As it continues to build, some sediment falls over the end. This side is called the leeward side. Grain flows occur when the windward side accumulates too much sediment, the angle of repose is reached and the sediment tumbles down. As more sediment piles on top the weight causes the underlying sediment to cement together and form cross beds.

Dish Structure

A dish structure is a type of sedimentary structure formed by liquefaction and fluidization of water-charged soft sediment either during or immediately following deposition. Dish structures are most commonly found in turbidites and other types of clastic deposits that result from subaqueous sediment gravity flows.

Terminology

Dish structure from Northern California

Due to the similarity in its shape to a dish, the structure, sometimes also referred to as dish-and-pillar or dish-and-pipe , was named after the common kitchen item.

History

Dish structure was described scientifically for the first time by Crook in 1961 who still used the title *discontinuous curved lamination*. The established term was used for the first time in 1967 by Stauffer and by Wentworth. Comprehensive studies are due to Lowe and LoPiccolo in 1974 and Lowe in 1975.

Description

[Nearly perfect dish in red, water escape indicated by yellow arrows. Upturned edge of a dish in blue.] The subhorizontal dish structure consists of two parts, the dish itself and the [sediment] contained within the dish plus the region stretching up to the bounding surface of the overlying dish(or dishes) above. The bounding surface of the dish can take on variable shapes, from substantially flat to bowl-like and to strongly concave up. The bounding surfaces are thin, (and usually) dark(er) laminae; they are richer in clay, silt or organic material than the surrounding sediment. The individual dishes are arranged *en echelon*. Their width can vary from 2 centimeters to over 50 centimeters, the vertical spacing ranges usually from less than 1 centimeter to about 8 centimeters. Their plan shape grades from circular/polygonal to oval/elliptical. Their bases are sharp, but the tops are gradational.

Commonly the dishes are separated by vertical streaks of massive sand called 'pillars'. These pillars can be small-scale structures (Type A pillars) or large and throughgoing high-discharge structures (Type B pillars).

Within an individual bed an increase in concavity combined with a simultaneous decrease in width of the dishes can often be observed towards the top.

Occurrence

Giant dish structure near Talara, Peru

Dish structure occurs in laterally extensive sheets. The medium in which the structure forms is usually coarse silt, but it also appears in all grades of sand. They are never found in gravels nor in clays. The containing beds are normally graded. The depositional environment of the structure is mainly deep-water marine (i.e. continental rise) comprising coarser grained turbidity currents and related high-concentration flows (grain flows, fluidized flows, liquefied flows). But dish structure can also be encountered in shallow-marine deposits and in fluviatile, lacustrine and delta environments. It is occasionally found in ash layers within marine sediments.

In turbidites dish structure usually forms within Bouma C, occasionally also within Bouma B.

Good examples of dish structure can be seen for instance in the Jack Fork Group in Oklahoma, in Ordovician turbidites at Cardigan in Wales, in deep-sea fan deposits near San Sebastián in Spain and in the Cerro Torro Formation of southern Chile. Some of the largest dish structure is found near Talara in northern Peru.

Formation

Up to 1974 dish structure was still regarded as a primary sedimentary structure. The formation of the structure was thought to be related either to the mechanics of sediment transport or to deposition in high-concentration gravity-flows. Only since Lowe and LoPiccolos's study, the structure is recognized as penecontemporaneous or secondary, formed during the dewatering of rapidly deposited quick or underconsolidated beds.

The postdepositional character of dish structure can sometimes clearly be seen in cut or displaced primary sedimentary structures (like convolute-laminated beds). During the dewatering process less permeable horizons (richer in small grain sizes like dispersed mud) act as barriers to upward flow; the flow is consequently forced sideways until an upward escape is possible. The sideways directed fluid motion has the tendency to leave fines along the low-permeability barriers which eventually become the clay-enriched laminae of the dishes. When the fluid finally finds a possibility to escape vertically it turns up the edges of the dishes. More forceful upward flow creates pillars – which are essentially dewatering pipes.

Use

Dish structure is a powerful means to recognize the younging direction in sediments.

Ripple Marks

Ancient wave ripple marks in sandstone, Moenkopi Formation, Capitol Reef National Park, Utah

Ripple beds in the Wren's Nest National Nature Reserve, Dudley, England

Cross-section through asymmetric climbing ripples, seen in the Zanskar Gorge, Ladakh, NW Indian Himalaya. Ripples climb when sediment fluxes in the flow are very high.

In geology, ripple marks are sedimentary structures (i.e. bedforms of the lower flow regime) and indicate agitation by water (current or waves) or wind.

Defining Ripple Cross-laminae and Asymmetric Ripples

Ripple marks in Cretaceous Dakota Formation, east side of Dinosaur Ridge. Scale bar on notebook is 10 cm.

- *Current ripple marks*, *unidirectional ripples*, or *asymmetrical ripple marks* are asymmetrical in profile, with a gentle up-current slope and a steeper down-current slope. The down-current slope is the angle of repose, which depends on the shape of the sediment. These commonly form in fluvial and aeolian depositional environments, and are a signifier of the lower part of the Lower Flow Regime.

- Ripple cross-laminae forms when deposition takes place during migration of current or wave ripples. A series of cross-laminae are produced by superimposing migrating ripples. The ripples form lateral to one another, such that the crests of vertically succeeding laminae are out of phase and appear to be advancing upslope. This process results in cross-bedded units that have the general appearance of waves in outcrop sections cut normal to the wave *crests*. In sections with other orientations, the laminae may appear horizontal or *trough-* shaped, depending upon the orientation and the shape of the ripples. Ripple cross-laminae will always have a steeper dip downstream, and will always be perpendicular to paleoflow meaning the orientation of the ripples will be in a direction that is ninety degrees to the direction that current if flowing. Scientists suggest current drag, or the slowing of current velocity, during deposition is responsible for ripple cross-laminae.

Types	
Straight	Straight ripples generate cross-laminae that all dip in the same direction, and lay in the same plane. These forms of ripples are constructed by unidirectional flow of the current.
Sinuous	Sinuous ripples generate cross-laminae that are curvy. They show a pattern of curving up and down as shown in picture. Sinuous ripples produce trough cross lamination. All laminae formed under this type of ripple dip at an angle to the flow as well as downstream. These are also formed by unidirectional flow of current.
Catenary	Catenary ripples generate cross-laminae that are curvy but have a unidirectional swoop. They show a pattern similar to what a repeated "W" would look like. Like the sinuous ripples, this form of ripple is created by unidirectional flow with the dip at an angle to the flow as well as downstream.

Linguoid / Lunate	
	Linguoid ripples have lee slope surfaces that are curved generating a laminae similar to caternary and sinuous ripples. Linguoid ripples generate an angle to the flow as well as downstream. Linguoid ripples have a random shape rather than a "W" shape, as described in the caternary description. Lunate ripples, meaning crescent shaped ripples, are exactly like linguoid ripples except that the *stoss* sides are curved rather than the *lee* slope. All other features are the same.

Size (scale)	
Size	**Description**
Very small	Very small cross-lamination means that the ripple height is roughly one centimeter. It is lenticular, wavy and flaser lamination.
Small	Small cross-bedding are ripples set at a height less than ten centimeters, while the thickness is only a few millimeters. Some ripples that may fit this category are wind ripples, wave ripples, and current ripples.
Medium	Medium cross-lamination are ripples with a height greater than ten centimeters, and less than one meter in thickness. Some ripples that may fit this category would be current-formed sandwaves, and storm-generated hummocky cross stratification.
Large	Large cross-bedding are ripples with a height greater than one meter, and a thickness equivalent to one meter or greater. Some ripples that may fit this category would be high energy river-bed bars, sand waves, epsilon cross-bedding and Gilbert-type cross-bedding.

Ripple marks in different environments

Wave-formed Ripples

- Also called bidirectional ripples, or symmetrical ripple marks have a symmetrical, almost sinusoidal profile; they indicate an environment with weak currents where water motion is dominated by wave oscillations.

- In most present-day streams, ripples will not form in sediment larger than coarse sand. Therefore, the stream beds of sand-bed streams are dominated by current ripples, while gravel-bed streams do not contain bedforms. The internal structure of ripples is a base of fine sand with coarse grains deposited on top since the size distribution of sand grains correlates to the size of the ripples. This occurs because the fine grains continue to move while the coarse grains accumulate and provide a protective barrier.

Ripple Marks Formed by Aeolian Processes

Wind ripples on crescent-shaped sand dunes (Barchans) in Southwest Afghanistan (Sistan).

Normal ripples

Also known as impact ripples, these occur in the lower part of the lower flow regime sands with grain sizes between 0.3-2.5 mm and normal ripples form wavelengths of 7-14 cm Normal ripples have straight or slightly sinuous crests approximately transverse to the direction of the wind.

Megaripples

These occur in the upper part of the lower flow regime where sand with bimodal particle size distribution forms unusually long wavelength of 1-25 m where the wind is not strong enough to move the larger particles but strong enough to move the smaller grains by saltation.

Fluid drag ripples

Also known as aerodynamic ripples, these are formed with fine, well-sorted grain particles accompanied by high velocity winds which result in long, flat ripples. The flat ripples are formed by long saltation paths taken by grains in suspension and grains on the ground surface.

Definitions

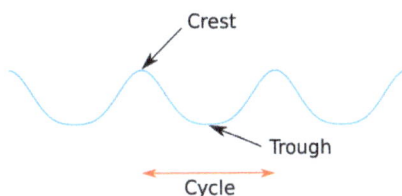

Crest and trough

Crest

> The point on a wave with the maximum value or height. It is the location at the peak of the wave cycle as shown in picture to the right.

Trough

> The opposite of a crest, so the minimum value or height in a wave. It is the location at the very lowest point of a wave cycle also shown in picture to right.

Lee

> The lee side has a steeper slope than the stoss. The lee is always on the back side of the ripple, which is also on the opposite side of where the current flow meets the ripple. The current flows down the lee side.

Stoss

> The stoss is the side of a wave or ripple that has a gentle slope versus a steeper slope. Current always flows up the stoss side and down the lee side. This can be used to determine current flow during the time of ripple formation.

Sole Markings

Sole marks are sedimentary structures found on the bases of certain strata, that indicate small-scale (usually on the order of centimetres) grooves or irregularities. This usually occurs at the interface of two differing lithologies and/or grain sizes. They are commonly preserved as casts of these indents on the bottom of the overlying bed (like flute casts). This is similar to casts and molds in fossil preservation. Occurring as they do only at the bottom of beds, and their distinctive shapes, they can make useful way up structures and paleocurrent indicators.

Sole markings are found most commonly in turbidite deposits, but are also often seen in modern river beds and tidal channels.

| Flute cast from the Book Cliffs of Utah | Load cast from drill core |

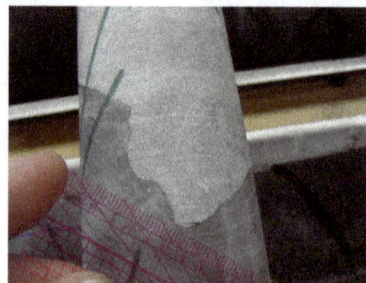

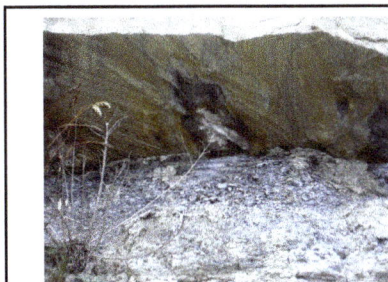

Groove casts on the base of a turbidite sandstone, Laga Basin, Italy

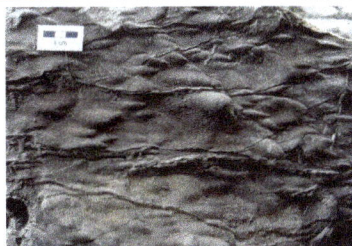

Flute casts on the base of a bed of sandstone from the Inverness Formation, western Cape Breton Island, Nova Scotia

History

Sole markings were first recognized in the Devonian rocks of New York State by James Hall in 1843. Originally, the features found on the undersides of beds were called *hieroglyphs*, *fucoids*, and *bio hieroglyphs*, because of their structure and how they were thought to be created; however, the term *sole mark* is used by geologists at present.

Scour Marks and Flute Casts

Scour marks and flute casts are scours dug into soft, fine sediment which typically get filled by an overlying bed (hence the name cast). Measuring the long axis of the flute cast gives the direction of flow, with the tapered end pointing toward the flow and the steep end up current. The concavity of the flute cast also points stratigraphically up. Flute casts can be characterized into four types, parabolic, spindle-shaped, comet-shaped, and asymmetrical.

- *Parabolic flute casts* are the most common and simple form. The shape of the bulbous end is parabolic or rounded shape and rarely shows any asymmetrical behavior. They occur in groups or individually, and all move parallel to each other and to paleoflow. In any given environment, their width and length will be consistent and range from a few centimeters, up to one meter in length.

- *Spindle-shaped flute casts* are found singularly or in groups, are considerably longer than they are wide and have pointed bulbous ends and are generally five to fifteen centimeters long. They can be quite shallow or as much as two thirds of the width in depth. These structures are easily definable because they lack symmetry parallel to the flow direction.

- *Comet-shaped flute casts* are characteristically found in isolation, and have a sharply pointed bulbous end, but the shallower end shows no stable continuous path. The overall length of the comet-shaped flute is rarely longer than ten centimeters, and the imprints are generally shallow.

- *Asymmetrical flute casts* are formed on top of a neighboring cast, and, therefore, covers half or more of the underlying flute. As the flutes continue to build outward in a step-like fashion and cut into each other, they get smaller and shallower.

Tool Marks

Tool marks are a type of sole marking formed by grooves left in a bed by things like sticks being dragged along by a current. The average direction of these can be assumed to be the flow direction, though it is bimodal, so it could be either way along the mark. Tool marks also have a more specific breakdown. There are grooves and striations, skip or prod marks, and roll marks. Groove or striation marks result from the continuous contact with the muddy bed. Skip or prod marks come from objects that bounce along the surface of the muddy bed. And roll marks result from objects rolling along the muddy bed.

- A *skip mark* is part of a series of linear tool marks left by an object that skipped along the bottom of a stream by saltation. Skip marks are characterized by their even spacing and the crescent-shaped mark that is left on the bed. The skip marks run parallel to paleoflow.

- *Saltation* is a method of sediment transport that briefly suspends particles and then drops them creating a forward bouncing pattern. This occurs because the turbulent currents are not strong enough to maintain suspension of the particles, but strong enough to suspend the particle for short bursts of time before the particle is returned to the sediment surface and bounces off again.

- A *prod mark* is a relatively short tool mark caused by an object that was dug into the muddy sediment and then lifted out. These markings are generally asymmetrical, getting deeper down current, and end suddenly.

- *Roll marks* are made by an object that was forced to roll down the bottom of a stream. The marks made in this case are continuous, long, generally linear, and run parallel to the paleoflow. The width dimensions of roll marks vary based upon the size of the object. Roll marks are a sign of water that has enough energy to cause motion but not enough turbidity and energy to separate the object from the bottom of a muddy bed.

Groove Casts

Groove casts are straight parallel ridges that are raised a few millimeters from the bedding surface. These structures were named and defined by Shrock in 1948 because of their long and narrow appearance, and they were formed from the filling in of grooves. Even though they may seem similar to flute casts, they each have many distinguishing characteristics, and the two are generally not found in the same vicinity. Groove casts are closely spaced, but not on top of each other, and exist in pairs, triples, and even larger groups. Groove casts form when high velocity flows (e.g. turbidite) create a pattern on an underlying bed. In 1957, Kuenen published that "groove cast" was a general term encompassing both drag marks and slide marks.

- A *slide mark* is a long, relatively wide, but shallow gouge left in a muddy bed caused by sliding of a soft-body object such as a bed of algae or slumping of sediment.

- *Drag marks* are narrower and deeper than slide marks, but retain the same length. Drag marks create a groove or striation caused by a physically hard object like a rock or shell.

Load Casts

Load casts are secondary structures that are preserved as bulbous depressions on the base of a bed. They form as dense, overlying sediment (usually sand) settles into less dense, water-saturated sediment (usually mud) below.

Soft-sediment Deformation Structures

Cross-sectional view of deformed beds caused by soft-sediment deformation in the Booti Booti Sandstone (Mississippian), New South Wales. (Rygel, M.C.)

Aztec Sandstone (Lower Jurassic) in southern Nevada showing distorted eolian sand beds.

Large soft sediment deformation structures in turbidites, SE Spain. These are probably best described as flame structures, or perhaps ball-and-pillow structures. Backpack is around half a meter high. A small reverse fault runs through the outcrop in the centre of the image.

Soft-sediment deformation structures develop at deposition or shortly after, during the first stages of the sediment's consolidation. This is because the sediments need to be "liquid-like" or unsolidified for the deformation to occur. These formations have also been put into a category called water-escape structures by Lowe (1975). The most common places for soft-sediment deformations to materialize are in deep water basins with turbidity currents, rivers, deltas, and shallow-marine areas with storm impacted conditions. This is because these environments have high deposition rates, which allows the sediments to pack loosely.

Types of Soft-sediment Deformation Structures

- Convolute bedding forms when complex folding and crumpling of beds or laminations occur. This type of deformation is found in fine or silty sands, and is usually confined to one rock layer. Convolute laminations are found in flood plain, delta, point-bar, and intertidal-flat deposits. They generally range in size from 3 to 25 cm, but there have been larger formations recorded as several meters thick.

- Flame structures consist of mud and are wavy or "flame" shaped. These flames usually extend into an overlying sandstone layer. This deformation is caused from sand being deposited onto mud, which is less dense. Load casts, technically a subset of sole markings, below, are the features which form alongside flame structures. Flames are thin fingers of mud injected upward into the overlying sands, while load casts are the pendulous knobs of sand that descend downwards into the mud between the flames.

- Slump structures are mainly found in sandy shales and mudstones, but may also be in limestones, sandstones, and evaporites. They are a result of the displacement and movement of unconsolidated sediments, and are found in areas with steep slopes and fast sedimentation rates. These structures often are faulted.

- Dish structures are thin, dish-shaped formations that normally occur in siltstones and sandstones. The size of each "dish" often ranges from 1 cm to 50 cm in size, and forms as a result of dewatering. Pillar structures often appear along with dish structures and also form by dewatering. They have a vertical orientation, which cuts across laminated or massive sands. These formations can range from a few millimeters in diameter to larger than a meter.

- Sole markings are found on the underside of sedimentary rocks that overlie shale beds, usually sandstones. They are used for determining the flow direction of old currents because of their directional features. Sole markings form from the erosion of a bed, which creates a groove that is later filled in by sediment.

- Seismites are sedimentary beds disturbed by seismic waves from earthquakes. They are commonly used to interpret the seismic history of an area. The term has also been applied to soft sediment deformation structures, including sand volcanos, sand blows, and certain clastic dikes.

References

- Boggs, S. (2005). Principles of Sedimentology and Stratigraphy (4th ed.). Upper Saddle River, N.J.: Prentice Hall. ISBN 0-13-099696-3.

- Stow, D.A.V. (2005). Sedimentary Rocks in the Field (1st ed.). Burlington, M.A.: Academic Press. ISBN 0-13-099696-3.

- Potter, P.E.; Maynard, J.B.; Depetris, P.J. (2005). Mud and Mudstones: Introduction and Overview (1st ed.). Wurzberg, Germany: Springer. ISBN 3-540-22157-3.

- Blatt, H.; Middleton, G.; Murray, R. (1980). Origin of Sedimentary Rocks (2nd ed.). Englewood Cliffs, N.J.: Prentice Hall. ISBN 0-13-642710-3.

- Schieber, J.; Zimmerle, W.; Sethi, P. (1998). Shales and Mudstones (1st ed.). Stuttgart, Germany: E. Schweizerbartsche Verlagsbuchhandlung. ISBN 3-510-65183-9.

- Tucker, M.E. (1994). Sedimentary Petrology: An Introduction to the Origin of Sedimentary Rocks (3rd ed.). Malden, M.A.: Blackwell. ISBN 0-632-05735-1.

- Nudds, J.R.; Selden, P.A. (2008). Fossil Ecosystems of North America: A Guide to the Sites and Their Extraordinary Biotas (1st ed.). Chicago: University Of Chicago Press. ISBN 0-226-60722-4.

- Prothero, Donald R. and Schwab, Fred (22 August 2003). Sedimentary Geology. Macmillan. pp. 265–269. ISBN 978-0-7167-3905-0. Retrieved 15 December 2012.

- Baturin, G.N, Phosphorites on the Sea Floor: Origin, Composition and Distribution. Elsevier. 1981, New York, pp. 24–50 ISBN 044441990X.

- Boggs, Sam, Jr. (2006). Principles of Sedimentology and Stratigraphy (4th ed.), Pearson Education Inc., Upper Saddle River, NJ, pp. 217–223 ISBN 0321643186

- Boggs, S. (2006) Principles of Sedimentology and Stratigraphy., 2nd ed. Printice Hall, New York. 662 pp. ISBN 0-13-154728-3

- Neuendorf, K.K.E., J.P. Mehl, Jr., and J.A. Jackson, eds. (2005) Glossary of Geology (5th ed.). Alexandria, Virginia, American Geological Institute. 779 pp. ISBN 0-922152-76-4

- Nichols, G. (2009) Sedimentology and Stratigraphy, 2nd ed. John Wiley & Sons Ltd, Chichester, West Sussex, United Kingdom. 419 pp. ISBN 978-1-4051-9379-5

- Tucker, M.E. (2003) Sedimentary Rocks in the Field, 3rd ed. John Wiley & Sons Ltd,West Sussex, England. 234 pp. ISBN 0-470-85123-6

- Flugel, E. (2010) Microfacies of Carbonate Rocks: Analysis, Interpretation and Application, 2nd ed. Springer-Verlag, Berlin, Germany. 984 pp. ISBN 978-3-642-03795-5

- Easterbrook, Don J. Surface processes and landforms. Upper Saddle River, N.J: Prentice Hall, 1999. Print. ISBN 0-13-860958-6 pp. 479-480.

- Monroe, James S., and Reed Wicander. The Changing Earth: Exploring Geology and Evolution, 2nd ed. Belmont: West Publishing Company, 1997. ISBN 0-314-09577-2 pp. 114-15, 352.

4

Processes Involved in Sedimentology

The processes involved in sedimentology are sediment transport, deposition, erosion, diagenesis, weathering etc. Deposition is the process in which sediments are added to a landmass whereas erosion is the process in which soil is removed from the surface of the Earth by wind or water flow and is transported to another location. This section serves as a source to understand the process involved in sedimentology.

Sediment Transport

Sediment transport is the movement of solid particles (sediment), typically due to a combination of gravity acting on the sediment, and/or the movement of the fluid in which the sediment is entrained. Sediment transport occurs in natural systems where the particles are clastic rocks (sand, gravel, boulders, etc.), mud, or clay; the fluid is air, water, or ice; and the force of gravity acts to move the particles along the sloping surface on which they are resting. Sediment transport due to fluid motion occurs in rivers, oceans, lakes, seas, and other bodies of water due to currents and tides. Transport is also caused by glaciers as they flow, and on terrestrial surfaces under the influence of wind. Sediment transport due only to gravity can occur on sloping surfaces in general, including hillslopes, scarps, cliffs, and the continental shelf—continental slope boundary.

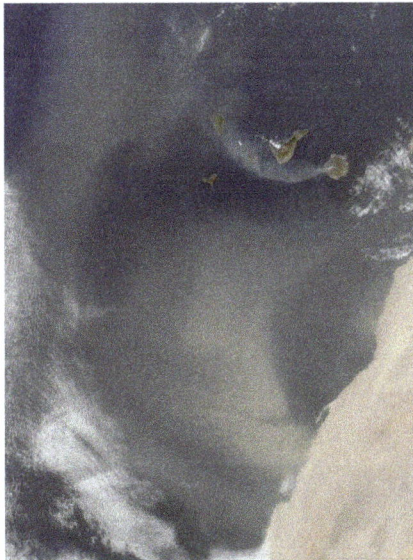

Dust blows from the Sahara Desert over the Atlantic Ocean towards the Canary Islands.

Sediment transport is important in the fields of sedimentary geology, geomorphology, civil engineering and environmental engineering. Knowledge of sediment transport is most often used to determine whether erosion or deposition will occur, the magnitude of this erosion or deposition, and the time and distance over which it will occur.

Mechanisms

Sand blowing off a crest in the Kelso Dunes of the Mojave Desert, California.

Toklat River, East Fork, Polychrome overlook, Denali National Park, Alaska. This river, like other braided streams, rapidly changes the positions of its channels through processes of erosion, sediment transport, and deposition.

Aeolian

Aeolian or *eolian* (depending on the parsing of æ) is the term for sediment transport by wind. This process results in the formation of ripples and sand dunes. Typically, the size of the transported sediment is fine sand (<1 mm) and smaller, because air is a fluid with low density and viscosity, and can therefore not exert very much shear on its bed.

Bedforms are generated by aeolian sediment transport in the terrestrial near-surface environment. Ripples and dunes form as a natural self-organizing response to sediment transport.

Aeolian sediment transport is common on beaches and in the arid regions of the world, because it is in these environments that vegetation does not prevent the presence and motion of fields of sand.

Wind-blown very fine-grained dust is capable of entering the upper atmosphere and moving across the globe. Dust from the Sahara deposits on the Canary Islands and islands in the Caribbean, and dust from the Gobi desert has deposited on the western United States. This sediment is important to the soil budget and ecology of several islands.

Deposits of fine-grained wind-blown glacial sediment are called loess.

Fluvial

In geology, physical geography, and sediment transport, fluvial processes relate to flowing water in natural systems. This encompasses rivers, streams, periglacial flows, flash floods and glacial lake outburst floods. Sediment moved by water can be larger than sediment moved by air because water has both a higher density and viscosity. In typical rivers the largest carried sediment is of sand and gravel size, but larger floods can carry cobbles and even boulders.

Fluvial sediment transport can result in the formation of ripples and dunes, in fractal-shaped patterns of erosion, in complex patterns of natural river systems, and in the development of floodplains.

Sand ripples, Laysan Beach, Hawaii. Coastal sediment transport results in these evenly spaced ripples along the shore. Monk seal for scale.

Coastal

Coastal sediment transport takes place in near-shore environments due to the motions of waves and currents. At the mouths of rivers, coastal sediment and fluvial sediment transport processes mesh to create river deltas.

Coastal sediment transport results in the formation of characteristic coastal landforms such as beaches, barrier islands, and capes.

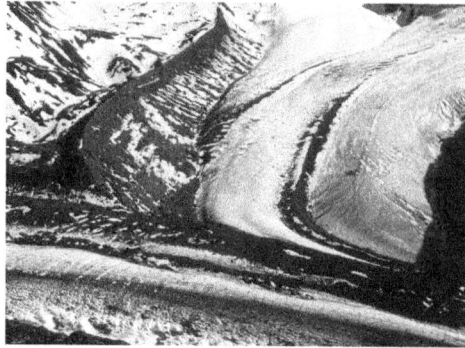

A glacier joining the Gorner Glacier, Zermatt, Switzerland. These glaciers transport sediment and leave behind lateral moraines.

Glacial

As glaciers move over their beds, they entrain and move material of all sizes. Glaciers can carry the largest sediment, and areas of glacial deposition often contain a large number of glacial erratics, many of which are several metres in diameter. Glaciers also pulverize rock into "glacial flour", which is so fine that it is often carried away by winds to create loess deposits thousands of kilometres afield. Sediment entrained in glaciers often moves approximately along the glacial flowlines, causing it to appear at the surface in the ablation zone.

Hillslope

In hillslope sediment transport, a variety of processes move regolith downslope. These include:

- Soil creep

- Tree throw

- Movement of soil by burrowing animals

- Slumping and landsliding of the hillslope

These processes generally combine to give the hillslope a profile that looks like a solution to the diffusion equation, where the diffusivity is a parameter that relates to the ease of sediment transport on the particular hillslope. For this reason, the tops of hills generally have a parabolic concave-up profile, which grades into a convex-up profile around valleys.

As hillslopes steepen, however, they become more prone to episodic landslides and other mass wasting events. Therefore, hillslope processes are better described by a nonlinear diffusion equation in which classic diffusion dominates for shallow slopes and erosion rates go to infinity as the hillslope reaches a critical angle of repose.

Debris Flow

Large masses of material are moved in debris flows, hyperconcentrated mixtures of mud, clasts that range up to boulder-size, and water. Debris flows move as granular flows down steep mountain valleys and washes. Because they transport sediment as a granular mixture, their transport mechanisms and capacities scale differently from those of fluvial systems.

Applications

Suspended sediment from a stream emptying into a fjord (Isfjorden, Svalbard, Norway).

Sediment transport is applied to solve many environmental, geotechnical, and geological problems. Measuring or quantifying sediment transport or erosion is therefore important for coastal engineering. Several sediment erosion devices have been designed in order to quantitfy sediment erosion (e.g., Particle Erosion Simulator (PES)). One such device, also referred to as the BEAST (Benthic Environmental Assessment Sediment Tool) has been calibrated in order to quantify rates of sediment erosion.

Movement of sediment is important in providing habitat for fish and other organisms in rivers. Therefore, managers of highly regulated rivers, which are often sediment-starved due to dams, are often advised to stage short floods to refresh the bed material and rebuild bars. This is also important, for example, in the Grand Canyon of the Colorado River, to rebuild shoreline habitats also used as campsites.

Sediment discharge into a reservoir formed by a dam forms a reservoir delta. This delta will fill the basin, and eventually, either the reservoir will need to be dredged or the dam will need to be removed. Knowledge of sediment transport can be used to properly plan to extend the life of a dam.

Geologists can use inverse solutions of transport relationships to understand flow depth, velocity, and direction, from sedimentary rocks and young deposits of alluvial materials.

Flow in culverts, over dams, and around bridge piers can cause erosion of the bed. This erosion can damage the environment and expose or unsettle the foundations of the structure. Therefore, good knowledge of the mechanics of sediment transport in a built environment are important for civil and hydraulic engineers.

When suspended sediment transport is increased due to human activities, causing environmental problems including the filling of channels, it is called siltation after the grain-size fraction dominating the process.

Initiation of Motion

Stress Balance

For a fluid to begin transporting sediment that is currently at rest on a surface, the boundary (or bed) shear stress τ_b exerted by the fluid must exceed the critical shear stress τ_c for the initiation of motion of grains at the bed. This basic criterion for the initiation of motion can be written as:

$$\tau_b = \tau_c.$$

This is typically represented by a comparison between a dimensionless shear stress ($\tau_b{}^*$) and a dimensionless critical shear stress ($\tau_c{}^*$). The nondimensionalization is in order to compare the driving forces of particle motion (shear stress) to the resisting forces that would make it stationary (particle density and size). This dimensionless shear stress, τ^*, is called the Shields parameter and is defined as:

$$\tau^* = \frac{\tau}{(\rho_s - \rho_f)(g)(D)}.$$

And the new equation to solve becomes:

$$\tau_b{}^* = \tau_c{}^*.$$

The equations included here describe sediment transport for clastic, or granular sediment. They do not work for clays and muds because these types of floccular sediments do not fit the geometric simplifications in these equations, and also interact thorough electrostatic forces. The equations were also designed for fluvial sediment transport of particles carried along in a liquid flow, such as that in a river, canal, or other open channel.

Only one size of particle is considered in this equation. However, river beds are often formed by a mixture of sediment of various sizes. In case of partial motion where only a part of the sediment mixture moves, the river bed becomes enriched in large gravel as the smaller sediments are washed away. The smaller sediments present under this layer of large gravel have a lower possibility of movement and total sediment transport decreases. This is called armouring effect. Other forms of armouring of sediment or decreasing rates of sediment erosion can be caused by carpets of microbial mats, under conditions of high organic loading.

Critical Shear Stress

Original Shields diagram, 1936

The Shields diagram empirically shows how the dimensionless critical shear stress (i.e. the dimensionless shear stress required for the initiation of motion) is a function of a particular form of the particle Reynolds number, Re_p or Reynolds number related to the particle. This allows us to rewrite the criterion for the initiation of motion in terms of only needing to solve for a specific version of the particle Reynolds number, which we call $\mathrm{Re}_p{}^*$.

$$\tau_b{}^* = f\left(\mathrm{Re}_p{}^*\right)$$

This equation can then be solved by using the empirically derived Shields curve to find $\tau_c{}^*$ as a function of a specific form of the particle Reynolds number called the boundary Reynolds number. The mathematical solution of the equation was given by Dey.

Particle Reynolds Number

In general, a particle Reynolds Number has the form:

$$\mathrm{Re}_p = \frac{U_p D}{\nu}$$

Where U_p is a characteristic particle velocity, D is the grain diameter (a characteristic particle size), and ν is the kinematic viscosity, which is given by the dynamic viscosity, μ, divided by the fluid density, ρ_f.

$$\nu = \frac{\mu}{\rho_f}$$

The specific particle Reynolds number of interest is called the boundary Reynolds number, and it is formed by replacing the velocity term in the Particle Reynolds number by the shear velocity, u_*, which is a way of rewriting shear stress in terms of velocity.

$$u_* = \sqrt{\frac{\tau_b}{\rho_f}} = \kappa z \frac{\partial u}{\partial z}$$

where τ_b is the bed shear stress (described below), and κ is the von Kármán constant, where

$$\kappa = 0.407.$$

The particle Reynolds number is therefore given by:

$$\mathrm{Re}_p{}^* = \frac{u_* D}{\nu}$$

Bed Shear Stress

The boundary Reynolds number can be used with the Shields diagram to empirically solve the equation

$$\tau_c{}^* = f\left(\mathrm{Re}_p{}^*\right),$$

which solves the right-hand side of the equation

$$\tau_b{}^* = \tau_c{}^*$$

In order to solve the left-hand side, expanded as

$$\tau_b{}^* = \frac{\tau_b}{(\rho_s - \rho_f)(g)(D)},$$

we must find the bed shear stress, τ_b. There are several ways to solve for the bed shear stress. First, we develop the simplest approach, in which the flow is assumed to be steady and uniform and reach-averaged depth and slope are used. Due to the difficulty of measuring shear stress *in situ*, this method is also one of the most-commonly used. This method is known as the depth-slope product.

Depth-slope Product

For a river undergoing approximately steady, uniform equilibrium flow, of approximately constant depth h and slope angle θ over the reach of interest, and whose width is much greater than its depth, the bed shear stress is given by some momentum considerations stating that the gravity force component in the flow direction equals exactly the friction force. For a wide channel, it yields:

$$\tau_b = \rho g h \sin(\theta)$$

For shallow slope angles, which are found in almost all natural lowland streams, the small-angle formula shows that $\sin(\theta)$ is approximately equal to $\tan(\theta)$, which is given by S, the slope. Rewritten with this:

$$\tau_b = \rho g h S$$

Shear Velocity, Velocity, and Friction Factor

For the steady case, by extrapolating the depth-slope product and the equation for shear velocity:

$$\tau_b = \rho g h S$$

$$u_* = \sqrt{\left(\frac{\tau_b}{\rho}\right)},$$

We can see that the depth-slope product can be rewritten as:

$$\tau_b = \rho u_*^2.$$

u_* is related to the mean flow velocity, $\bar{u}$, through the generalized Darcy-Weisbach friction factor, C_f, which is equal to the Darcy-Weisbach friction factor divided by 8 (for mathematical convenience). Inserting this friction factor,

$$\tau_b = \rho C_f \left(\bar{u}\right)^2$$

Unsteady Flow

For all flows that cannot be simplified as a single-slope infinite channel (as in the depth-slope product, above), the bed shear stress can be locally found by applying the Saint-Vennant equations for continuity, which consider accelerations within the flow.

Example

Set-up

The criterion for the initiation of motion, established earlier, states that

$$\tau_b{}^* = \tau_c{}^*.$$

In this equation,

$$\tau^* = \frac{\tau}{(\rho_s - \rho)(g)(D)}, \quad \text{and therefore}$$

$$\frac{\tau_b}{(\rho_s - \rho)(g)(D)} = \frac{\tau_c}{(\rho_s - \rho)(g)(D)}.$$

$\tau_c{}^*$ is a function of boundary Reynolds number, a specific type of particle Reynolds number.

$$\tau_c{}^* = f\left(Re_p{}^*\right).$$

For a particular particle Reynolds number, $\tau_c{}^*$ will be an emprical constant given by the Shields Curve or by another set of empirical data (depending on whether or not the grain size is uniform).

Therefore, the final equation that we seek to solve is:

$$\frac{\tau_b}{(\rho_s - \rho)(g)(D)} = f\left(Re{}^*\right).$$

Solution

We make several assumptions to provide an example that will allow us to bring the above form of the equation into a solved form.

First, we assume that the a good approximation of reach-averaged shear stress is given by the depth-slope product. We can then rewrite the equation as

$$\rho g h S = 0.06(\rho_s - \rho)(g)(D)..$$

Moving and re-combining the terms, we obtain:

$$hS = \frac{(\rho_s - \rho)}{\rho}(D)\left(f\left(Re_p{}^*\right)\right) = RD\left(f\left(Re_p{}^*\right)\right)$$

where R is the submerged specific gravity of the sediment.

We then make our second assumption, which is that the particle Reynolds number is high. This is typically applicable to particles of gravel-size or larger in a stream, and means that the critical shear stress is a constant. The Shields curve shows that for a bed with a uniform grain size,

$$\tau_c{}^* = 0.06.$$

Later researchers have shown that this value is closer to

$$\tau_c{}^* = 0.03$$

for more uniformly sorted beds. Therefore, we will simply insert

$$\tau_c{}^* = f\left(Re_p{}^*\right)$$

and insert both values at the end.

The equation now reads:

$$hS = RD\tau_c *$$

This final expression shows that the product of the channel depth and slope is equal to the Shield's criterion times the submerged specific gravity of the particles times the particle diameter.

For a typical situation, such as quartz-rich sediment $\left(\rho_s = 2650\dfrac{kg}{m^3}\right)$ in water $\left(\rho = 1000\dfrac{kg}{m^3}\right)$, the submerged specific gravity is equal to 1.65.

$$R = \frac{(\rho_s - \rho)}{\rho} = 1.65$$

Plugging this into the equation above,

$$hS = 1.65(D)\tau_c *.$$

For the Shield's criterion of $\tau_c * = 0.06.$. 0.06 * 1.65 = 0.099, which is well within standard margins of error of 0.1. Therefore, for a uniform bed,

$$hS = 0.1(D).$$

For these situations, the product of the depth and slope of the flow should be 10% of the diameter of the median grain diameter.

The mixed-grain-size bed value is $\tau_c * = 0.03$, which is supported by more recent research as being more broadly applicable because most natural streams have mixed grain sizes. Using this value, and changing D to D_50 ("50" for the 50th percentile, or the median grain size, as we are now looking at a mixed-grain-size bed), the equation becomes:

$$hS = 0.05(D_{50})$$

Which means that the depth times the slope should be about 5% of the median grain diameter in the case of a mixed-grain-size bed.

Modes of Entrainment

The sediments entrained in a flow can be transported along the bed as bed load in the form of sliding and rolling grains, or in suspension as suspended load advected by the main flow. Some sediment materials may also come from the upstream reaches and be carried downstream in the form of wash load.

Rouse Number

The location in the flow in which a particle is entrained is determined by the Rouse number, which is determined by the density ρ_s and diameter d of the sediment particle, and the density ρ and kinematic viscosity v of the fluid, determine in which part of the flow the sediment particle will be carried.

$$P = \frac{w_s}{\kappa u_*}$$

Here, the Rouse number is given by P. The term in the numerator is the (downwards) sediment the sediment settling velocity w_s, which is discussed below. The upwards velocity on the grain is given as a product of the von Kármán constant, $\kappa = 0.4$, and the shear velocity, u_*.

The following table gives the approximate required Rouse numbers for transport as bed load, suspended load, and wash load.

Mode of Transport	Rouse Number
Initiation of motion	>7.5
Bed load	>2.5, <7.5
Suspended load: 50% Suspended	>1.2, <2.5
Suspended load: 100% Suspended	>0.8, <1.2
Wash load	<0.8

Settling Velocity

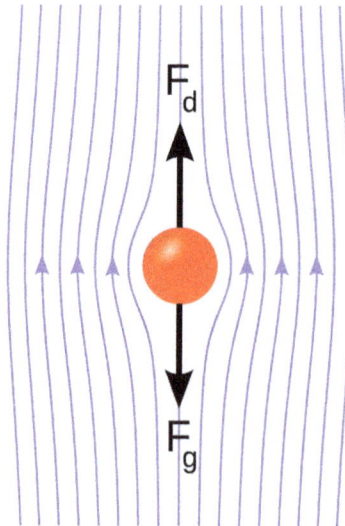

Streamlines around a sphere falling through a fluid. This illustration is accurate for laminar flow, in which the particle Reynolds number is small. This is typical for small particles falling through a viscous fluid; larger particles would result in the creation of a turbulent wake.

The settling velocity (also called the "fall velocity" or "terminal velocity") is a function of the particle Reynolds number. Generally, for small particles (laminar approximation), it can be calculated with Stokes' Law. For larger particles (turbulent particle Reynolds numbers), fall velocity is calculated with the turbulent drag law. Dietrich (1982) compiled a large amount of published data to which he empirically fit settling velocity curves. Ferguson and Church (2006) analytically combined the expressions for Stokes flow and a turbulent drag law into a single equation that works for all sizes of sediment, and successfully tested it against the data of Dietrich. Their equation is

$$w_s = \frac{RgD^2}{C_1 v + (0.75 C_2 R g D^3)^{(0.5)}}.$$

In this equation w_s is the sediment settling velocity, g is acceleration due to gravity, and D is mean sediment diameter. v is the kinematic viscosity of water, which is approximately 1.0×10^{-6} m²/s for water at 20 °C.

C_1 and C_2 are constants related to the shape and smoothness of the grains.

Con-stant	Smooth Spheres	Natural Grains: Sieve Diameters	Natural Grains: Nominal Diameters	Limit for Ultra-Angular Grains
	18	18	20	24
	0.4	1.0	1.1	1.2

The expression for fall velocity can be simplified so that it can be solved only in terms of D. We use the sieve diameters for natural grains, $g = 9.8,$, and values given above for v and R. From these parameters, the fall velocity is given by the expression:

$$w_s = \frac{16.17 D^2}{1.8 \cdot 10^{-5} + (12.1275 D^3)^{(0.5)}}$$

Hjulström-Sundborg Diagram

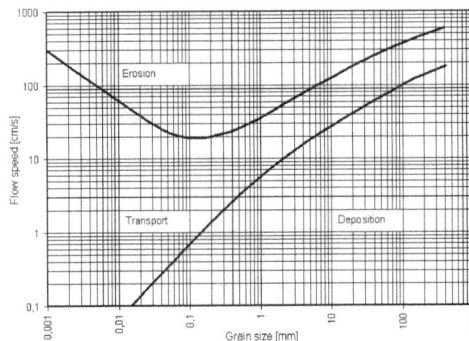

The logarithmic Hjulström curve

In 1935, Filip Hjulström created the Hjulström curve, a graph which shows the relationship between the size of sediment and the velocity required to erode (lift it), transport it, or deposit it. The graph is logarithmic.

Åke Sundborg later modified the Hjulström curve to show separate curves for the movement threshold corresponding to several water depths, as is necessary if the flow velocity rather than the boundary shear stress (as in the Shields diagram) is used for the flow strength.

This curve has no more than a historical value nowadays, although its simplicity is still attractive. Among the drawbacks of this curve are that it does not take the water depth into account and more importantly, that it does not show that sedimentation is caused by flow velocity *deceleration* and erosion is caused by flow *acceleration*. The dimensionless Shields diagram is now unanimously accepted for initiation of sediment motion in rivers. Much work was done on river sediment transport formulae in the second half of the 20th century and that work should be used preferably to Hjulström's curve, e.g. Meyer-Peter & Müller (1948), Engelund-Hansen (1967), Lefort (1991), Belleudy (2012).

Transport Rate

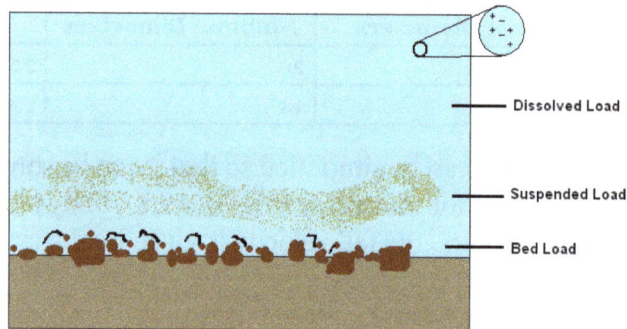

A schematic diagram of where the different types of sediment load are carried in the flow. Dissolved load is not sediment: it is composed of disassociated ions moving along with the flow. It may, however, constitute a significant proportion (often several percent, but occasionally greater than half) of the total amount of material being transported by the stream.

Formulas to calculate sediment transport rate exist for sediment moving in several different parts of the flow. These formulas are often segregated into bed load, suspended load, and wash load. They may sometimes also be segregated into bed material load and wash load.

Bed Load

Bed load moves by rolling, sliding, and hopping (or saltating) over the bed, and moves at a small fraction of the fluid flow velocity. Bed load is generally thought to constitute 5-10% of the total sediment load in a stream, making it less important in terms of mass balance. However, the bed material load (the bed load plus the portion of the suspended load which comprises material derived from the bed) is often dominated by bed load, especially in gravel-bed rivers. This bed material load is the only part of the sediment load that actively interacts with the bed. As the bed load is an important component of that, it plays a major role in controlling the morphology of the channel.

Bed load transport rates are usually expressed as being related to excess dimensionless shear stress raised to some power. Excess dimensionless shear stress is a nondimensional measure of bed shear stress about the threshold for motion.

$$(\tau_b^* - \tau_c^*),$$

Bed load transport rates may also be given by a ratio of bed shear stress to critical shear stress, which is equivalent in both the dimensional and nondimensional cases. This ratio is called the "transport stage" (T_s or ϕ) and is an important in that it shows bed shear stress as a multiple of the value of the criterion for the initiation of motion.

$$T_s = \phi = \frac{\tau_b}{\tau_c}$$

When used for sediment transport formulae, this ratio is typically raised to a power.

The majority of the published relations for bedload transport are given in dry sediment weight per unit channel width, b ("breadth"):

$$q_s = \frac{Q_s}{b}.$$

Due to the difficulty of estimating bed load transport rates, these equations are typically only suitable for the situations for which they were designed.

Notable Bed Load Transport Formulae

Meyer-Peter Müller and Derivatives

The transport formula of Meyer-Peter and Müller, originally developed in 1948, was designed for well-sorted fine gravel at a transport stage of about 8. The formula uses the above nondimensionalization for shear stress,

$$\tau^* = \frac{\tau}{(\rho_s - \rho)(g)(D)},$$

and Hans Einstein's nondimensionalization for sediment volumetric discharge per unit width

$$q_s^* = \frac{q_s}{D\sqrt{\dfrac{\rho_s - \rho}{\rho}gD}} = \frac{q_s}{Re_p v}.$$

Their formula reads:

$$q_s^* = 8\left(\tau^* - \tau^*_c\right)^{3/2}.$$

Their experimentally determined value for τ^*_c is 0.047, and is the third commonly used value for this (in addition to Parker's 0.03 and Shields' 0.06).

Because of its broad use, some revisions to the formula have taken place over the years that show that the coefficient on the left ("8" above) is a function of the transport stage:

$$T_s \approx 2 \rightarrow q_s^* = 5.7\left(\tau^* - 0.047\right)^{3/2}$$

$$T_s \approx 100 \rightarrow q_s^* = 12.1\left(\tau^* - 0.047\right)^{3/2}$$

The variations in the coefficient were later generalized as a function of dimensionless shear stress:

$$\begin{cases} q_s^* = \alpha_s \left(\tau^* - \tau_c^*\right)^n \\ n = \dfrac{3}{2} \\ \alpha_s = 1.6\ln\left(\tau^*\right) + 9.8 \approx 9.64\tau^{*0.166} \end{cases}$$

Wilcock and Crowe

In 2003, Peter Wilcock and Joanna Crowe (now Joanna Curran) published a sediment transport formula that works with multiple grain sizes across the sand and gravel range. Their formula works with surface grain size distributions, as opposed to older models which use subsurface grain size distributions (and thereby implicitly infer a surface grain sorting).

Their expression is more complicated than the basic sediment transport rules (such as that of Meyer-Peter and Müller) because it takes into account multiple grain sizes: this requires consideration of reference shear stresses for each grain size, the fraction of the total sediment supply that falls into each grain size class, and a "hiding function".

The "hiding function" takes into account the fact that, while small grains are inherently more mobile than large grains, on a mixed-grain-size bed, they may be trapped in deep pockets between large grains. Likewise, a large grain on a bed of small particles will be stuck in a much smaller pocket than if it were on a bed of grains of the same size. In gravel-bed rivers, this can cause "equal mobility", in which small grains can move just as easily as large ones. As sand is added to the system, it moves away from the "equal mobility" portion of the hiding function to one in which grain size again matters.

Their model is based on the transport stage, or ratio of bed shear stress to critical shear stress for the initiation of grain motion. Because their formula works with several grain sizes simultaneously, they define the critical shear stress for each grain size class, τ_{c,D_i}, to be equal to a "reference shear stress", τ_{ri}. .

They express their equations in terms of a dimensionless transport parameter, W_i^* (with the "$*$" indicating nondimensionality and the "i" indicating that it is a function of grain size):

$$W_i^* = \frac{Rgq_{bi}}{F_i u^{*3}}$$

q_{bi} is the volumetric bed load transport rate of size class i per unit channel width b. F_i is the proportion of size class i that is present on the bed.

They came up with two equations, depending on the transport stage, ϕ. For $\phi < 1.35$:

$$W_i^* = 0.002\phi^{7.5}$$

and for $\phi \geq 1.35$:

$$W_i^* = 14\left(1 - \frac{0.894}{\phi^{0.5}}\right)^{4.5}.$$

This equation asymptotically reaches a constant value of W_i^* as ϕ becomes large.

Wilcock and Kenworthy

In 2002, Peter Wilcock and Kenworthy T.A., following Peter Wilcock (1998), published a sediment bed-load transport formula that works with only two sediments fractions, i.e. sand and gravel fractions. Peter Wilcock and Kenworthy T.A. in their article recognized that a mixed-sized sediment bed-load transport model using only two fractions offers practical advantages in terms of both computational and conceptual modeling by taking into account the nonlinear effects of sand presence in gravel beds on bed-load transport rate of both fractions. In fact, in the two-fraction bed load formula appears a new ingredient with respect to that of Meyer-Peter and Müller that is the proportion F_i of fraction i on the bed surface where the subscript i represents either the sand (s) or gravel (g) fraction. The proportion F_i, as a function of sand content f_s, physically represents the relative influence of the mechanisms controlling sand and gravel transport, associated with the change from a clast-supported to matrix-supported gravel bed. Moreover, since f_s spans between 0 and 1, phenomena that vary with f_s include the relative size effects producing "hiding" of fine grains and "exposure" of coarse grains. The "hiding" effect takes into account the fact that, while small grains are inherently more mobile than large grains, on a mixed-grain-size bed, they may be trapped in deep pockets between large grains. Likewise, a large grain on a bed of small particles will be stuck in a much smaller pocket than if it were on a bed of grains of the same size, which the Meyer-Peter and Müller formula refers to. In gravel-bed rivers, this can cause "equal mobility", in which small grains can move just as easily as large ones. As sand is added to the system, it moves away from the "equal mobility" portion of the hiding function to one in which grain size again matters.

Their model is based on the transport stage, i.e. ϕ, or ratio of bed shear stress to critical shear stress for the initiation of grain motion. Because their formula works with only two fractions simultaneously, they define the critical shear stress for each of the two grain size classes, τ_{ri}, where $_i$ represents either the sand (s) or gravel (g) fraction. The critical shear stress that represents the incipient motion for each of the two fractions is consistent with established values in the limit of pure sand and gravel beds and shows a sharp change with increasing sand content over the transition from a clast- to matrix-supported bed.

They express their equations in terms of a dimensionless transport parameter, W_i^* (with the "*" indicating nondimensionality and the "$_i$" indicating that it is a function of grain size):

$$W_i^* = \frac{Rgq_{bi}}{F_i u^{*3}}$$

q_{bi} is the volumetric bed load transport rate of size class i per unit channel width b. F_i is the proportion of size class i that is present on the bed.

They came up with two equations, depending on the transport stage, ϕ. For $\phi < \phi$:

$$W_i^* = 0.002\phi^{7.5}$$

and for $\phi \geq \phi$:

$$W_i^* = A\left(1 - \frac{\chi}{\phi^{0.5}}\right)^{4.5}$$

This equation asymptotically reaches a constant value of W_i^* as ϕ becomes large and the symbols A, ϕ', χ have the following values:

$$A = 70, \phi' = 1.19, \chi = 0.908, laboratory$$

$$A = 115, \phi' = 1.27, \chi = 0.923, field$$

In order to apply the above formulation, it is necessary to specify the characteristic grain sizes D_s for the sand portion and D_g for the gravel portion of the surface layer, the fractions F_s and F_g of sand and gravel, respectively in the surface layer, the submerged specific gravity of the sediment R and shear velocity associated with skin friction u_*.

Kuhnle et al.

For the case in which sand fraction is transported by the current over and through an immobile gravel bed, Kuhnle et al.(2013), following the theoretical analysis done by Pellachini (2011), provides a new relationship for the bed load transport of the sand fraction when gravel particles remain at rest. It is worth mentioning that Kuhnle et al. (2013) applied the Wilcock and Kenworthy (2002) formula to their experimental data and found out that predicted bed load rates of sand fraction were about 10 times greater than measured and approached 1 as the sand elevation became near the top of the gravel layer. They, also, hypothesized that the mismatch between predicted and measured sand bed load rates is due to the fact that the bed shear stress used for the Wilcock and Kenworthy (2002) formula was larger than that available for transport within the gravel bed because of the sheltering effect of the gravel particles. To overcome this mismatch, following Pellachini (2011), they assumed that the variability of the bed shear stress available for the sand to be transported by the current would be some function of the so-called "Roughness Geometry Function" (RGF), which represents the gravel bed elevations distribution. Therefore, the sand bed load formula follows as:

$$q_s^* = 2.29*10^{-5} A(z_s)^{2.14} \left(\frac{\tau_b}{\tau_{cs}} \right)^{3.49}$$

where

$$q_s^* = \frac{q_s}{[(s-1)gD_s]^{0.5} \rho_s D_s}$$

the subscript $_s$ refers to the sand fraction, s represents the ratio ρ_s / ρ_w where ρ_s is the sand fraction density, $A(z_s)$ is the RGF as a function of the sand level z_s within the gravel bed, τ_b is the bed shear stress available for sand transport and τ_{cs} is the critical shear stress for incipient motion of the sand fraction, which was calculated graphically using the updated Shields-type relation of Miller et al.(1977).

Suspended Load

Suspended load is carried in the lower to middle parts of the flow, and moves at a large fraction of the mean flow velocity in the stream.

A common characterization of suspended sediment concentration in a flow is given by the Rouse Profile. This characterization works for the situation in which sediment concentration c_0 at one particular elevation above the bed z_0 can be quantified. It is given by the expression:

$$\frac{c_s}{c_0} = \left[\frac{z(h-z_0)}{z_0(h-z)} \right]^{-P/\alpha}$$

Here, z is the elevation above the bed, c_s is the concentration of suspended sediment at that elevation, h is the flow depth, P is the Rouse number, and α relates the eddy viscosity for momentum K_m to the eddy diffusivity for sediment, which is approximately equal to one.

$$\alpha = \frac{K_s}{K_m} \approx 1$$

Experimental work has shown that α ranges from 0.93 to 1.10 for sands and silts.

The Rouse profile characterizes sediment concentrations because the Rouse number includes both turbulent mixing and settling under the weight of the particles. Turbulent mixing results in the net motion of particles from regions of high concentrations to low concentrations. Because particles settle downward, for all cases where the particles are not neutrally buoyant or sufficiently light that this settling velocity is negligible, there is a net negative concentration gradient as one goes upward in the flow. The Rouse Profile therefore gives the concentration profile that provides a balance between turbulent mixing (net upwards) of sediment and the downwards settling velocity of each particle.

Bed Material Load

Bed material load comprises the bed load and the portion of the suspended load that is sourced from the bed.

Three common bed material transport relations are the "Ackers-White", "Engelund-Hansen", "Yang" formulae. The first is for sand to granule-size gravel, and the second and third are for sand though Yang later expanded his formula to include fine gravel. That all of these formulae cover the sand-size range and two of them are exclusively for sand is that the sediment in sand-bed rivers is commonly moved simultaneously as bed and suspended load.

Engelund-Hansen

The bed material load formula of Engelund and Hansen is the only one to not include some kind of critical value for the initiation of sediment transport. It reads:

$$q_s* = \frac{0.05}{c_f} \tau*^{2.5}$$

where q_s* is the Einstein nondimensionalization for sediment volumetric discharge per unit width, c_f is a friction factor, and $\tau*$ is the Shields stress. The Engelund-Hansen formula is one of the few sediment transport formulae in which a threshold "critical shear stress" is absent.

Wash Load

Wash load is carried within the water column as part of the flow, and therefore moves with the mean velocity of main stream. Wash load concentrations are approximately uniform in the water column. This is described by the endmember case in which the Rouse number is equal to 0 (i.e. the settling velocity is far less than the turbulent mixing velocity), which leads to a prediction of a perfectly uniform vertical concentration profile of material.

Total Load

Some authors have attempted formulations for the total sediment load carried in water. These formulas are designed largely for sand, as (depending on flow conditions) sand often can be carried as both bed load and suspended load in the same stream or shoreface.

Deposition (Geology)

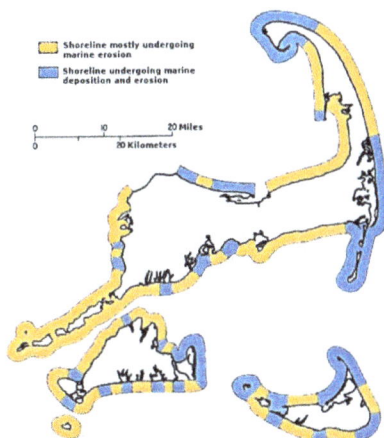

Map of Cape Cod showing shores undergoing erosion (cliffed sections) in yellow, and shores characterized by marine deposition (barriers) in blue.

Deposition is the geological process in which sediments, soil and rocks are added to a landform or land mass. Wind, ice, and water, as well as sediment flowing via gravity, transport previously eroded sediment, which, at the loss of enough kinetic energy in the fluid, is deposited, building up layers of sediment.

Deposition occurs when the forces responsible for sediment transportation are no longer sufficient to overcome the forces of gravity and friction, creating a resistance to motion, this is known as the null-point hypothesis. Deposition can also refer to the buildup of sediment from organically derived matter or chemical processes. For example, chalk is made up partly of the microscopic calcium carbonate skeletons of marine plankton, the deposition of which has induced chemical processes (diagenesis) to deposit further calcium carbonate. Similarly, the formation of coal begins with deposition of organic material, mainly from plants, in anaerobic conditions.

Null-point Hypothesis

The null-point hypothesis explains how sediment is deposited throughout a shore profile according to its grain size. This is due to the influence of hydraulic energy, resulting in a seaward-fining of sediment particle size, or where fluid forcing equals gravity for each grain size. The concept can also be explained as "sediment of a particular size may move across the profile to a position where it is in equilibrium with the wave and flows acting on that sediment grain". This sorting mechanism combines the influence of the down-slope gravitational force of the profile and forces due to flow asymmetry, the position where there is zero net transport is known as the null point and was first proposed by Cornaglia in 1889. Figure 1 illustrates this relationship between sediment grain size and the depth of the marine environment.

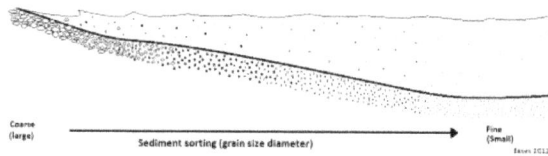

Figure 1. Illustrates the sediment size distribution over a shoreline profile, where finer sediments are transported away from high energy environments and settle out of suspension, or deposit in calmer environments. Coarse sediments are maintained in the upper shoreline profile and are sorted by the wave-generated hydraulic regime

The first principle underlying the null point theory is due to the gravitational force; finer sediments remain in the water column for longer durations allowing transportation outside the surf zone to deposit under calmer conditions. The gravitational effect, or settling velocity determines the location of deposition for finer sediments, whereas a grain's internal angle of friction determines the deposition of larger grains on a shore profile. The secondary principle to the creation of seaward sediment fining is known as the hypothesis of asymmetrical thresholds under waves; this describes the interaction between the oscillatory flow of waves and tides flowing over the wave ripple bedforms in an asymmetric pattern. "The relatively strong onshore stroke of the wave forms an eddy or vortex on the lee side of the ripple, provided the onshore flow persists, this eddy remains trapped in the lee of the ripple. When the flow reverses, the eddy is thrown upwards off the bottom and a small cloud of suspended sediment generated by the eddy is ejected into the water column above the ripple, the sediment cloud is then moved seaward by the offshore stroke of the wave." Where there is symmetry in ripple shape the vortex is neutralised, the eddy and its associated sediment cloud develops on both sides of the ripple. This creates a cloudy water column which travels under tidal influence as the wave orbital motion is in equilibrium.

The Null-point hypothesis has been quantitatively proven in Akaroa Harbour, New Zealand, The Wash, U.K., Bohai Bay and West Huang Sera, Mainland China, and in numerous other studies; Ippen and Eagleson (1955), Eagleson and Dean (1959, 1961) and Miller and Zeigler (1958, 1964).

Deposition of Non-cohesive Sediments

Large grain sediments transported by either bed load or suspended load will come to rest when there is insufficient bed shear stress and fluid turbulence to keep the sediment moving, with the suspended load this can be some distance as the particles need to fall through the water column. This is determined by the grains downward acting weight force being matched by a combined buoyancy and fluid drag force and can be expressed by:

$$\frac{4}{3}\pi R^3 \rho_s g = \frac{4}{3}\pi R^3 \rho g + \frac{1}{2}C_d \rho \pi R^2 w_s^2$$

Downward acting weight force = Upward-acting buoyancy force + Upward-acting fluid drag force

where:

- π is the ratio of a circle's circumference to its diameter.

- R is the radius of the spherical object (in m),

- ρ is the mass density of the fluid (kg/m³),

- g is the gravitational acceleration (m/s²),

- C_d is the drag coefficient, and

- w_s is the particle's settling velocity (in m/s).

In order to calculate the drag coefficient, the grain's Reynolds number needs to be discovered, which is based on the type of fluid through which the sediment particle is flowing; laminar flow, turbulent flow or a hybrid of both. When the fluid becomes more viscous due to smaller grain sizes or larger settling velocities, prediction is less straight forward and it is applicable to incorporate Stokes Law (also known as the frictional force, or drag force) of settling.

Deposition of Cohesive Sediments

Cohesion of sediment occurs with the small grain sizes associated with silts and clays, or particles smaller than 4□ on the phi scale. If these fine particles remain dispersed in the water column, Stokes law applies to the settling velocity of the individual grains, although due to sea water being a strong electrolyte bonding agent, flocculation occurs where individual particles create an electrical bond adhering each other together to form flocs. "The face of a clay platelet has a slight negative charge where the edge has a slight positive charge, when two platelets come into close proximity with each other the face of one particle and the edge of the other are electrostatically attracted." Flocs then have a higher combined mass which leads to quicker deposition through a higher fall velocity, and deposition in a more shoreward direction than they would have as the individual fine grains of clay or silt.

The Occurrence of Null Point Theory

Akaroa Harbour is located on Banks Peninsula, Canterbury, New Zealand,

43°48'S 172°56'E43.800°S 172.933°E. The formation of this harbour has occurred due to active erosional processes on an extinct shield volcano, whereby the sea has flooded the caldera creating an inlet 16 km in length, with an average width of 2 km and a depth of -13 m relative to mean sea level at the 9 km point down the transect of the central axis. The predominant storm wave energy has unlimited fetch for the outer harbour from a southerly direction, with a calmer environment within the inner harbour, though localised harbour breezes create surface currents and chop influencing the marine sedimentation processes. Deposits of loess from subsequent glacial periods have in filled volcanic fissures over millennia, resulting in volcanic basalt and loess as the main sediment types available for deposition in Akaroa Harbour

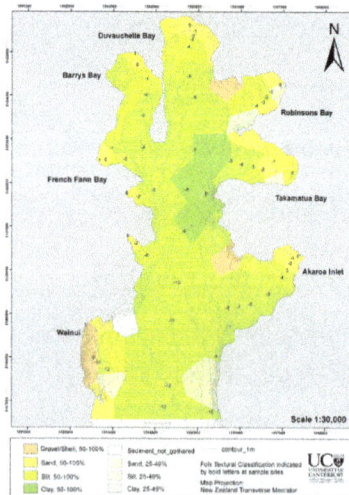

Figure. Map of Akaroa Harbour showing a fining of sediments with increased bathymetry toward the central axis of the harbour. Taken from Hart et al. (2009) and the University of Canterbury under contract of Environment Canterbury.

Hart et al. (2009) discovered through bathymetric survey, sieve and pipette analysis of subtidal sediments, that sediment textures were related to three main factors: depth; distance from shoreline; and distance along the central axis of the harbour. Resulting in the fining of sediment textures with increasing depth and towards the central axis of the harbour, or if classified into grain class sizes, "the plotted transect for the central axis goes from silty sands in the intertidal zone, to sandy silts in the inner nearshore, to silts in the outer reaches of the bays to mud at depths of 6 m or more"

Other studies have shown this process of the winnowing of sediment grain size from the effect of hydrodynamic forcing; Wang, Collins and Zhu (1988) qualitatively correlated increasing intensity of fluid forcing with increasing grain size. "This correlation was demonstrated at the low energy clayey tidal flats of Bohai Bay (China), the moderate environment of the Jiangsu coast (China) where the bottom material is silty, and the sandy flats of the high energy coast of The Wash (U.K.)." This research shows conclusive evidence for the null point theory existing on tidal flats with differing hydrodynamic energy levels and also on flats that are both erosional and accretional.

Kirby R. (2002) takes this concept further explaining that the fines are suspended and reworked aerially offshore leaving behind lag deposits of mainly bivalve and gastropod shells separated out from the finer substrate beneath, waves and currents then heap these deposits to form chenier ridges throughout the tidal zone which tend to be forced up the foreshore profile but also along the foreshore. Cheniers can be found at any level on the foreshore and predominantly characterise an erosion-dominated regime.

Applications for Coastal Planning and Management

The null point theory has been controversial in its acceptance into mainstream coastal science as the theory operates in dynamic equilibrium or unstable equilibrium, and many field and laboratory observations have failed to replicate the state of a null point at each grain size throughout the profile. The interaction of variables and processes over time within the environmental context causes issues; "the large number of variables, the complexity of the processes, and the difficulty in observation, all place serious obstacles in the way of systematisation, therefore in certain narrow fields the basic physical theory may be sound and reliable but the gaps are large"

Geomorphologists, engineers, governments and planners should be aware of the processes and outcomes involved with the null point hypothesis when performing tasks such as beach nourishment, issuing building consents or building coastal defence structures. This is because sediment grain size analysis throughout a profile allows inference into the erosion or accretion rates possible if shore dynamics are modified. Planners and managers should also be aware that the coastal environment is dynamic and contextual science should be evaluated before implementation of any shore profile modification. Thus theoretical studies, laboratory experiments, numerical and hydraulic modelling seek to answer questions pertaining to littoral drift and sediment deposition, the results should not be viewed in isolation and a substantial body of purely qualitative observational data should supplement any planning or management decision.

Erosion

An actively eroding rill on an intensively-farmed field in eastern Germany

In earth science, erosion is the action of surface processes (such as water flow or wind) that remove soil, rock, or dissolved material from one location on the Earth's crust, then transport it away to another location. The particulate breakdown of rock or soil into clastic sediment is referred to as *physical* or *mechanical* erosion; this contrasts with *chemical* erosion, where soil or rock material is removed from an area by its dissolving into a solvent (typically water), followed by the flow away of that solution. Eroded sediment or solutes may be transported just a few millimetres, or for thousands of kilometres.

Natural rates of erosion are controlled by the action of geomorphic drivers, such as rainfall; bedrock wear in rivers; coastal erosion by the sea and waves; glacial plucking, abrasion, and scour; areal flooding; wind abrasion; groundwater processes; and mass movement processes in steep landscapes like landslides and debris flows. The rates at which such processes act control how fast a surface is eroded. Typically, physical erosion proceeds fastest on steeply sloping surfaces, and rates may also be sensitive to some climatically-controlled properties including amounts of water supplied (e.g., by rain), storminess, wind speed, wave fetch, or atmospheric temperature (especially for some ice-related processes). Feedbacks are also possible between rates of erosion and the amount of eroded material that is already carried by, for example, a river or glacier. Processes of erosion that produce sediment or solutes from a place contrast with those of deposition, which control the arrival and emplacement of material at a new location.

While erosion is a natural process, human activities have increased by 10-40 times the rate at which erosion is occurring globally. Excessive (or accelerated) erosion causes both "on-site" and "off-site" problems. On-site impacts include decreases in agricultural productivity and (on natural landscapes) ecological collapse, both because of loss of the nutrient-rich upper soil layers. In some cases, the eventual end result is desertification. Off-site effects include sedimentation of waterways and eutrophication of water bodies, as well as sediment-related damage to roads and houses. Water and wind erosion are the two primary causes of land degradation; combined, they are responsible for about 84% of the global extent of degraded land, making excessive erosion one of the most significant environmental problems worldwide.

Intensive agriculture, deforestation, roads, anthropogenic climate change and urban sprawl are amongst the most significant human activities in regard to their effect on stimulating erosion. However, there are many prevention and remediation practices that can curtail or limit erosion of vulnerable soils.

A natural arch produced by the wind erosion of differentially weathered rock in Jebel Kharaz, Jordan.

A wave-like sea cliff produced by coastal erosion, in Jinshitan Coastal National Geopark, Dalian, Liaoning Province, China.

Physical Processes

Rainfall and Surface Runoff

Soil and water being splashed by the impact of a single raindrop.

Rainfall, and the surface runoff which may result from rainfall, produces four main types of soil erosion: *splash erosion, sheet erosion, rill erosion,* and *gully erosion.* Splash erosion is generally seen as the first and least severe stage in the soil erosion process, which is followed by sheet erosion, then rill erosion and finally gully erosion (the most severe of the four).

In *splash erosion,* the impact of a falling raindrop creates a small crater in the soil, ejecting soil particles. The distance these soil particles travel can be as much as 0.6 m (two feet) vertically and 1.5 m (five feet) horizontally on level ground.

If the soil is saturated, or if the rainfall rate is greater than the rate at which water can infiltrate into the soil, surface runoff occurs. If the runoff has sufficient flow energy, it will transport loosened soil particles (sediment) down the slope. *Sheet erosion* is the transport of loosened soil particles by overland flow.

A spoil tip covered in rills and gullies due to erosion processes caused by rainfall: Rummu, Estonia

Rill erosion refers to the development of small, ephemeral concentrated flow paths which function as both sediment source and sediment delivery systems for erosion on hillslopes. Generally, where water erosion rates on disturbed upland areas are greatest, rills are active. Flow depths in rills are typically of the order of a few centimetres (about an inch) or less and along-channel slopes may be quite steep. This means that rills exhibit hydraulic physics very different from water flowing through the deeper, wider channels of streams and rivers.

Gully erosion occurs when runoff water accumulates and rapidly flows in narrow channels during or immediately after heavy rains or melting snow, removing soil to a considerable depth.

Rivers and Streams

Valley or *stream erosion* occurs with continued water flow along a linear feature. The erosion is both downward, deepening the valley, and headward, extending the valley into the hillside, creating head cuts and steep banks. In the earliest stage of stream erosion, the erosive activity is dominantly vertical, the valleys have a typical V cross-section and the stream gradient is relatively steep. When some base level is reached, the erosive activity switches to lateral erosion, which widens the valley floor and creates a narrow floodplain. The stream gradient becomes nearly flat, and lateral deposition of sediments becomes important as the stream meanders across the valley floor. In all stages of stream erosion, by far the most erosion occurs during times of flood, when more and faster-moving water is available to carry a larger sediment load. In such processes, it is not the water alone that erodes: suspended abrasive particles, pebbles and boulders can also act erosively as they traverse a surface, in a process known as *traction*.

Dobbingstone Burn, Scotland, showing two different types of erosion affecting the same place. Valley erosion is occurring due to the flow of the stream, and the boulders and stones (and much of the soil) that are lying on the stream's banks are glacial till that was left behind as ice age glaciers flowed over the terrain.

Bank erosion is the wearing away of the banks of a stream or river. This is distinguished from changes on the bed of the watercourse, which is referred to as *scour*. Erosion and changes in the form of river banks may be measured by inserting metal rods into the bank and marking the position of the bank surface along the rods at different times.

Thermal erosion is the result of melting and weakening permafrost due to moving water. It can occur both along rivers and at the coast. Rapid river channel migration observed in the Lena River of Siberia is due to thermal erosion, as these portions of the banks are composed of permafrost-cemented non-cohesive materials. Much of this erosion occurs as the weakened banks fail in large slumps. Thermal erosion also affects the Arctic coast, where wave action and near-shore temperatures combine to undercut permafrost bluffs along the shoreline and cause them to fail. Annual erosion rates along a 100-kilometre (62-mile) segment of the Beaufort Sea shoreline averaged 5.6 metres (18 feet) per year from 1955 to 2002.

Coastal Erosion

Wave cut platform caused by erosion of cliffs by the sea, at Southerndown in South Wales.

Erosion of the boulder clay (of Pleistocene age) along cliffs of Filey Bay, Yorkshire, England.

Shoreline erosion, which occurs on both exposed and sheltered coasts, primarily occurs through the action of currents and waves but sea level (tidal) change can also play a role.

Hydraulic action takes place when air in a joint is suddenly compressed by a wave closing the entrance of the joint. This then cracks it. *Wave pounding* is when the sheer energy of the wave hitting the cliff or rock breaks pieces off. *Abrasion* or *corrasion* is caused by waves launching seaload at the cliff. It is the most effective and rapid form of shoreline erosion. *Corrosion* is the dissolving of rock by carbonic acid in sea water. Limestone cliffs are particularly vulnerable to this kind of erosion. *Attrition* is where particles/seaload carried by the waves are worn down as they hit each other and the cliffs. This then makes the material easier to wash away. The material ends up as shingle and sand. Another significant source of erosion, particularly on carbonate coastlines, is the boring, scraping and grinding of organisms, a process termed *bioerosion*.

Sediment is transported along the coast in the direction of the prevailing current (longshore drift). When the upcurrent amount of sediment is less than the amount being carried away, erosion occurs. When the upcurrent amount of sediment is greater, sand or gravel banks will tend to form as a result of deposition. These banks may slowly migrate along the coast in the direction of the longshore drift, alternately protecting and exposing parts of the coastline. Where there is a bend in the coastline, quite often a buildup of eroded material occurs forming a long narrow bank (a spit). Armoured beaches and submerged offshore sandbanks may also protect parts of a coastline from erosion. Over the years, as the shoals gradually shift, the erosion may be redirected to attack different parts of the shore.

Chemical Erosion

Chemical erosion is the loss of matter in a landscape in the form of solutes. Chemical erosion is usually calculated from the solutes found in streams. Anders Rapp pioneered the study of chemical erosion in his work about Kärkevagge published in 1960.

Glaciers

Glacial moraines above Lake Louise, in Alberta, Canada.

Glaciers erode predominantly by three different processes: abrasion/scouring, plucking, and ice thrusting. In an abrasion process, debris in the basal ice scrapes along the bed, polishing and gouging the underlying rocks, similar to sandpaper on wood. Glaciers can also cause pieces of bedrock to crack off in the process of plucking. In ice thrusting, the glacier freezes to its bed, then as it surges forward, it moves large sheets of frozen sediment at the base along with the glacier. This method produced some of the many thousands of lake basins that dot the edge of the Canadian Shield. The erosion caused by glaciers worldwide erodes mountains so effectively that the term *glacial buzz-saw* has become widely used, which describes the limiting effect of glaciers on the height of mountain ranges. As mountains grow higher, they generally allow for more glacial activity (especially in the accumulation zone above the glacial equilibrium line altitude), which causes increased rates of erosion of the mountain, decreasing mass faster than isostatic rebound can add to the mountain. This provides a good example of a negative feedback loop. Ongoing research is showing that while glaciers tend to decrease mountain size, in some areas, glaciers can actually reduce the rate of erosion, acting as a *glacial armour*.

These processes, combined with erosion and transport by the water network beneath the glacier, leave moraines, drumlins, ground moraine (till), kames, kame deltas, moulins, and glacial erratics in their wake, typically at the terminus or during glacier retreat.

Floods

At extremely high flows, kolks, or vortices are formed by large volumes of rapidly rushing water. Kolks cause extreme local erosion, plucking bedrock and creating pothole-type geographical features called Rock-cut basins. Examples can be seen in the flood regions result from glacial Lake Missoula, which created the channeled scablands in the Columbia Basin region of eastern Washington.

Wind Erosion

Wind erosion is a major geomorphological force, especially in arid and semi-arid regions. It is also a major source of land degradation, evaporation, desertification, harmful airborne dust, and crop damage—especially after being increased far above natural rates by human activities such as deforestation, urbanization, and agriculture.

Árbol de Piedra, a rock formation in the Altiplano, Bolivia sculpted by wind erosion.

Wind erosion is of two primary varieties: *deflation*, where the wind picks up and carries away loose particles; and *abrasion*, where surfaces are worn down as they are struck by airborne particles carried by wind. Deflation is divided into three categories: (1) *surface creep*, where larger, heavier particles slide or roll along the ground; (2) *saltation*, where particles are lifted a short height into the air, and bounce and saltate across the surface of the soil; and (3) *suspension*, where very small and light particles are lifted into the air by the wind, and are often carried for long distances. Saltation is responsible for the majority (50-70%) of wind erosion, followed by suspension (30-40%), and then surface creep (5-25%).

Wind erosion is much more severe in arid areas and during times of drought. For example, in the Great Plains, it is estimated that soil loss due to wind erosion can be as much as 6100 times greater in drought years than in wet years.

Mass Movement

Wadi in Makhtesh Ramon, Israel, showing gravity collapse erosion on its banks.

Mass movement is the downward and outward movement of rock and sediments on a sloped surface, mainly due to the force of gravity.

Mass movement is an important part of the erosional process, and is often the first stage in the breakdown and transport of weathered materials in mountainous areas. It moves material from higher elevations to lower elevations where other eroding agents such as streams and glaciers can then pick up the material and move it to even lower elevations. Mass-movement processes are always occurring continuously on all slopes; some mass-movement processes act very slowly; others occur very suddenly, often with disastrous results. Any perceptible down-slope movement of rock or sediment is often referred to in general terms as a landslide. However, landslides can be classified in a much more detailed way that reflects the mechanisms responsible for the movement and the velocity at which the movement occurs. One of the visible topographical manifestations of a very slow form of such activity is a scree slope.

Slumping happens on steep hillsides, occurring along distinct fracture zones, often within materials like clay that, once released, may move quite rapidly downhill. They will often show a spoon-shaped isostatic depression, in which the material has begun to slide downhill. In some cases, the slump is caused by water beneath the slope weakening it. In many cases it is simply the result of poor engineering along highways where it is a regular occurrence. *Surface creep* is the slow movement of soil and rock debris by gravity which is usually not perceptible except through extended observation. However, the term can also describe the rolling of dislodged soil particles 0.5 to 1.0 mm (0.02 to 0.04 in) in diameter by wind along the soil surface.

Factors Affecting Erosion Rates

Climate

The amount and intensity of precipitation is the main climatic factor governing soil erosion by water. The relationship is particularly strong if heavy rainfall occurs at times when, or in locations where, the soil's surface is not well protected by vegetation. This might be during periods when agricultural activities leave the soil bare, or in semi-arid regions where vegetation is naturally sparse. Wind erosion requires strong winds, particularly during times of drought when vegetation is sparse and soil is dry (and so is more erodible). Other climatic factors such as average temperature and temperature range may also affect erosion, via their effects on vegetation and soil properties. In general, given similar vegetation and ecosystems, areas with more precipitation (especially high-intensity rainfall), more wind, or more storms are expected to have more erosion.

In some areas of the world (e.g. the mid-western USA), rainfall intensity is the primary determinant of erosivity, with higher intensity rainfall generally resulting in more soil erosion by water. The size and velocity of rain drops is also an important factor. Larger and higher-velocity rain drops have greater kinetic energy, and thus their impact will displace soil particles by larger distances than smaller, slower-moving rain drops.

In other regions of the world (e.g. western Europe), runoff and erosion result from relatively low intensities of stratiform rainfall falling onto previously saturated soil. In such situations, rainfall amount rather than intensity is the main factor determining the severity of soil erosion by water.

Vegetative Cover

Vegetation acts as an interface between the atmosphere and the soil. It increases the permeability of the soil to rainwater, thus decreasing runoff. It shelters the soil from winds, which results in decreased wind erosion, as well as advantageous changes in microclimate. The roots of the plants bind the soil together, and interweave with other roots, forming a more solid mass that is less susceptible to both water and wind erosion. The removal of vegetation increases the rate of surface erosion.

Topography

The topography of the land determines the velocity at which surface runoff will flow, which in turn determines the erosivity of the runoff. Longer, steeper slopes (especially those without adequate vegetative cover) are more susceptible to very high rates of erosion during heavy rains than shorter, less steep slopes. Steeper terrain is also more prone to mudslides, landslides, and other forms of gravitational erosion processes.

Tectonics

Tectonic processes control rates and distributions of erosion at the Earth's surface. If tectonic action causes part of the Earth's surface (e.g., a mountain range) to be raised or lowered relative to surrounding areas, this must necessarily change the gradient of the land surface. Because erosion rates are almost always sensitive to local slope, this will change the rates of erosion in the uplifted area. Active tectonics also brings fresh, unweathered rock towards the surface, where it is exposed to the action of erosion.

However, erosion can also affect tectonic processes. The removal by erosion of large amounts of rock from a particular region, and its deposition elsewhere, can result in a lightening of the load on the lower crust and mantle. Because tectonic processes are driven by gradients in the stress field developed in the crust, this unloading can in turn cause tectonic or isostatic uplift in the region. In some cases, it has been hypothesised that these twin feedbacks can act to localise and enhance zones of very rapid exhumation of deep crustal rocks beneath places on the Earth's surface with extremely high erosion rates, for example, beneath the extremely steep terrain of Nanga Parbat in the western Himalayas. Such a place has been called a "tectonic aneurysm".

Erosion of Earth Systems

Mountain Ranges

Mountain ranges are known to take many millions of years to erode to the degree they effectively cease to exist. Scholars Pitman and Golovchenko estimate that it takes probably more than 450 million years to erode a mountain mass similar to the Himalaya into an almost-flat peneplain if there are no major sea-level changes. Erosion of mountains massifs can create a pattern of equally high summits called summit accordance. It has been argued that extension during post-orogenic collapse is a more effective mechanism of lowering the height of orogenic mountains than erosion.

Examples of heavily eroded mountain ranges include the Timanides of Northern Russia. Erosion of this orogen has produced sediments that are now found in the East European Platform, including the Cambrian Sablya Formation near Lake Ladoga. Studies of these sediments points that its likely that the erosion of the orogen was beginning in the Cambrian and then became stronger in Ordovician.

Soils

If the rate of erosion is higher than the rate of soil formation the soils are being destroyed by erosion. Where soil is not destroyed by erosion, erosion can in some cases prevent the formation of soil features that form slowly. Inceptisols are common soils that form in areas of fast erosion.

While erosion of soils is a natural process, human activities have increased by 10-40 times the rate at which erosion is occurring globally. Excessive (or accelerated) erosion causes both "on-site" and "off-site" problems. On-site impacts include decreases in agricultural productivity and (on natural landscapes) ecological collapse, both because of loss of the nutrient-rich upper soil layers. In some cases, the eventual end result is desertification. Off-site effects include sedimentation of waterways and eutrophication of water bodies, as well as sediment-related damage to roads and houses. Water and wind erosion are the two primary causes of land degradation; combined, they are responsible for about 84% of the global extent of degraded land, making excessive erosion one of the most significant environmental problems worldwide.

Diagenesis

Diagenesis is the change of sediments or existing sedimentary rocks into a different sedimentary rock during and after rock formation (lithification), at temperatures and pressures less than that required for the formation of metamorphic rocks. It does not include changes from weathering. It is any chemical, physical, or biological change undergone by a sediment after its initial deposition, after its lithification. This process excludes surface alteration (weathering) and metamorphism. These changes happen at relatively low temperatures and pressures and result in changes to the rock's original mineralogy and texture. There is no sharp boundary between diagenesis and metamorphism, but the latter occurs at higher temperatures and pressures than the former.

Hydrothermal solutions, meteoric groundwater, porosity, permeability, solubility, and time are all influential factors.

After deposition, sediments are compacted as they are buried beneath successive layers of sediment and cemented by minerals that precipitate from solution. Grains of sediment, rock fragments and fossils can be replaced by other minerals during diagenesis. Porosity usually decreases during diagenesis, except in rare cases such as dissolution of minerals and dolomitization.

The study of diagenesis in rocks is used to understand the geologic history they have undergone and the nature and type of fluids that have circulated through them. From a commercial standpoint, such studies aid in assessing the likelihood of finding various economically viable mineral and hydrocarbon deposits.

The process of diagenesis is also important in the decomposition of bone tissue.

The Role of Diagenesis in Anthropology and Paleontology

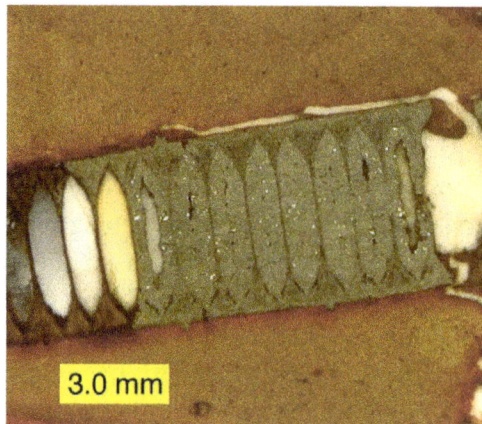

Originally calcitic crinoid stem (in cross-section) diagenetically replaced by marcasite in a siderite concretion; Lower Carboniferous.

The term diagenesis, literally meaning "across generation", is extensively used in geology. However, this term has filtered into the field of anthropology, archaeology and paleontology to describe the changes and alterations that take place on skeletal (biological) material. Specifically, diagenesis "is the cumulative physical, chemical and biological environment; these processes will modify an organic object's original chemical and/or structural properties and will govern its ultimate fate, in terms of preservation or destruction". In order to assess the potential impact of diagenesis on archaeological or fossil bones, many factors need to be assessed, beginning with elemental and mineralogical composition of bone and enveloping soil, as well as the local burial environment (geology, climatology, groundwater).

The composite nature of bone, comprising one-third organic (mainly protein collagen) and two thirds mineral (calcium phosphate mostly in the form of hydroxyapatite) renders its diagenesis more complex. Alteration occurs at all scales from molecular loss and substitution, through crystallite reorganization, porosity and microstructural changes, and in many cases, to disintegration of the complete unit. Three general pathways of the diagenesis of bone have been identified:

1. chemical deterioration of the organic phase.

2. chemical deterioration of the mineral phase.

3. (micro) biological attack of the composite.

They are as follows:

1. The dissolution of collagen depends on time, temperature and environmental pH. At high temperatures, the rate of collagen loss will be accelerated and extreme pH can cause collagen swelling and accelerated hydrolysis. Due to the increase in porosity of bones through collagen loss, the bone becomes susceptible to hydrolytic infiltration where the hydroxyapatite, with its affinity for amino acids, permits charged species of endogenous and exogenous origin to take up residence.

2. The hydrolytic activity plays a key role in the mineral phase transformations that exposes the collagen to accelerated chemical- and bio-degradation. Chemical changes affect crystallinity. Mechanisms of chemical change, such as the uptake of F^- or CO_3^- may cause recrystallization where hydroxyapatite is dissolved and re-precipitated allowing for the incorporation of substitution of exogenous material.

3. Once an individual has been interred, microbial attack, the most common mechanism of bone deterioration, occurs rapidly. During this phase, most bone collagen is lost and porosity is increased. The dissolution of the mineral phase caused by low pH permits access to the collagen by extracellular microbial enzymes thus microbial attack.

The Role of Diagenesis in Hydrocarbon Generation

When animal or plant matter is buried during sedimentation, the constituent organic molecules (lipids, proteins, carbohydrates and lignin-humic compounds) break down due to the increase in temperature and pressure. This transformation occurs in the first few hundred meters of burial and results in the creation of two primary products: kerogens and bitumens.

It is generally accepted that hydrocarbons are formed by the thermal alteration of these kerogens (the *biogenic* theory). In this way, given certain conditions (which are largely temperature-dependent) kerogens will break down to form hydrocarbons through a chemical process known as cracking, or catagenesis.

A kinetic model based on experimental data can capture most of the essential transformation in diagenesis, and a mathematical model in a compacting porous medium to model the dissolution-precipitation mechanism. These models have been intensively studied and applied in real geological applications.

Diagenesis has been divided, based on hydrocarbon and coal genesis into: *eodiagenesis* (early), *mesodiagenesis* (middle) and *telodiagenesis* (late). During the early or eodiagenesis stage shales lose pore water, little to no hydrocarbons are formed and coal varies between lignite and sub-bituminous. During mesodiagenesis, dehydration of clay minerals occurs, the main development of oil genesis occurs and high to low volatile bituminous coals are formed. During telodiagenesis organic matter undergoes cracking and dry gas is produced; semi-anthracite coals develop.

Early diagenesis in newly formed aquatic sediments is mediated by microorganisms using different electron acceptors as part of their metabolism. Organic matter is mineralized, liberating gaseous carbon dioxide (CO_2) in the porewater, which, depending on the conditions, can diffuse into the water column. The various processes of mineralization in this phase are nitrification and denitrification, manganese oxide reduction, iron hydroxide reduction, sulfate reduction, and fermentation.

The Role of Diagenesis in Bone Decomposition

Diagenesis alters the proportions of organic collagen and inorganic components (hydroxyapatite, calcium, magnesium) of bone exposed to environmental conditions, especially moisture. This is accomplished by the exchange of natural bone constituents, deposition in voids or defects, adsorption onto the bone surface and leaching from the bone.

Weathering

Weathering is the breaking down of rocks, soil, and minerals as well as wood and artificial materials through contact with the Earth's atmosphere, waters, and biological organisms. Weathering occurs *in situ* (on site), that is, in the same place, with little or no movement, and thus should not be confused with erosion, which involves the movement of rocks and minerals by agents such as water, ice, snow, wind, waves and gravity and then being transported and deposited in other locations.

Two important classifications of weathering processes exist – physical and chemical weathering; each sometimes involves a biological component. Mechanical or physical weathering involves the breakdown of rocks and soils through direct contact with atmospheric conditions, such as heat, water, ice and pressure. The second classification, chemical weathering, involves the direct effect of atmospheric chemicals or biologically produced chemicals also known as biological weathering in the breakdown of rocks, soils and minerals. While physical weathering is accentuated in very cold or very dry environments, chemical reactions are most intense where the climate is wet and hot. However, both types of weathering occur together, and each tends to accelerate the other. For example, physical abrasion (rubbing together) decreases the size of particles and therefore increases their surface area, making them more susceptible to rapid chemical reactions. The various agents act in concert to convert primary minerals (feldspars and micas) to secondary minerals (clays and carbonates) and release plant nutrient elements in soluble forms.

The materials left over after the rock breaks down combined with organic material creates soil. The mineral content of the soil is determined by the parent material; thus, a soil derived from a single rock type can often be deficient in one or more minerals needed for good fertility, while a soil weathered from a mix of rock types (as in glacial, aeolian or alluvial sediments) often makes more fertile soil. In addition, many of Earth's landforms and landscapes are the result of weathering processes combined with erosion and re-deposition.

Physical Weathering

Physical weathering, also recognized as mechanical weathering, is the class of processes that causes the disintegration of rocks without chemical change. The primary process in physical weathering is abrasion (the process by which clasts and other particles are reduced in size). However, chemical and physical weathering often go hand in hand. Physical weathering can occur due to temperature, pressure, frost etc. For example, cracks exploited by physical weathering will increase the surface area exposed to chemical action, thus amplifying the rate of disintegration.

Abrasion by water, ice, and wind processes loaded with sediment can have tremendous cutting power, as is amply demonstrated by the gorges, ravines, and valleys around the world. In glacial areas, huge moving ice masses embedded with soil and rock fragments grind down rocks in their path and carry away large volumes of material. Plant roots sometimes enter cracks in rocks and pry them apart, resulting in some disintegration; Burrowing animals may help disintegrate rock through their physical action. However, such influences are usually of little importance in producing parent material when compared to the drastic physical effects of water, ice, wind, and temperature change. Physical weathering is also called mechanical weathering or disaggregation.

Thermal Stress

Thermal stress weathering (sometimes called insolation weathering) results from the expansion and contraction of rock, caused by temperature changes. For example, heating of rocks by sunlight or fires can cause expansion of their constituent minerals. As some minerals expand more than others, temperature changes set up differential stresses that eventually cause the rock to crack apart. Because the outer surface of a rock is often warmer or colder than the more protected inner portions, some rocks may weather by exfoliation – the peeling away of outer layers. This process may be sharply accelerated if ice forms in the surface cracks. When water freezes, it expands with a force of about 1465 Mg/m^2, disintegrating huge rock masses and dislodging mineral grains from smaller fragments.

Thermal stress weathering comprises two main types, thermal shock and thermal fatigue. Thermal stress weathering is an important mechanism in deserts, where there is a large diurnal temperature range, hot in the day and cold at night. The repeated heating and cooling exerts stress on the outer layers of rocks, which can cause their outer layers to peel off in thin sheets. The process of peeling off is also called exfoliation. Although temperature changes are the principal driver, moisture can enhance thermal expansion in rock. Forest fires and range fires are also known to cause significant weathering of rocks and boulders exposed along the ground surface. Intense localized heat can rapidly expand a boulder.

The thermal heat from wildfire can cause significant weathering of rocks and boulders, heat can rapidly expand a boulder and thermal shock can occur. The differential expansion of a thermal gradient can be understood in terms of stress or of strain, equivalently. At some point, this stress can exceed the strength of the material, causing a crack to form. If nothing stops this crack from propagating through the material, it will result in the object's structure to fail.

Frost Weathering

A rock in Abisko, Sweden fractured along existing joints possibly by frost weathering or thermal stress

Frost weathering, frost wedging, *ice wedging* or *cryofracturing* is the collective name for several processes where ice is present. These processes include frost shattering, frost-wedging and freeze-thaw weathering. Severe frost shattering produces huge piles of rock fragments called scree which may be located at the foot of mountain areas or along slopes. Frost weathering is common in mountain areas where the temperature is around the freezing point of water. Certain frost-susceptible soils expand or heave upon freezing as a result of water migrating via capillary action to grow ice lenses near the freezing front. This same phenomenon occurs within pore spaces of rocks. The ice accumulations grow larger as they attract liquid water from the surrounding pores. The ice crystal growth weakens the rocks which, in time, break up. It is caused by the approximately 10% (9.87) expansion of ice when water freezes, which can place considerable stress on anything containing the water as it freezes.

Freeze induced weathering action occurs mainly in environments where there is a lot of moisture, and temperatures frequently fluctuate above and below freezing point, especially in alpine and periglacial areas. An example of rocks susceptible to frost action is chalk, which has many pore spaces for the growth of ice crystals. This process can be seen in Dartmoor where it results in the formation of tors. When water that has entered the joints freezes, the ice formed strains the walls of the joints and causes the joints to deepen and widen. When the ice thaws, water can flow further into the rock. Repeated freeze-thaw cycles weaken the rocks which, over time, break up along the joints into angular pieces. The angular rock fragments gather at the foot of the slope to form a talus slope (or scree slope). The splitting of rocks along the joints into blocks is called block disintegration. The blocks of rocks that are detached are of various shapes depending on rock structure.

Ocean Waves

Wave action and water chemistry lead to structural failure in exposed rocks

Coastal geography is formed by the weathering of wave actions over geological times or can happen more abruptly through the process of salt weathering.

Pressure Release

In pressure release, also known as unloading, overlying materials (not necessarily rocks) are removed (by erosion, or other processes), which causes underlying rocks to expand and fracture parallel to the surface.

Pressure release could have caused the exfoliated granite sheets shown in the picture.

Intrusive igneous rocks (e.g. granite) are formed deep beneath the Earth's surface. They are under tremendous pressure because of the overlying rock material. When erosion removes the overlying rock material, these intrusive rocks are exposed and the pressure on them is released. The outer parts of the rocks then tend to expand. The expansion sets up stresses which cause fractures parallel to the rock surface to form. Over time, sheets of rock break away from the exposed rocks along the fractures, a process known as exfoliation. Exfoliation due to pressure release is also known as "sheeting".

Retreat of an overlying glacier can also lead to exfoliation due to pressure release.

Salt-crystal Growth

Tafoni at Salt Point State Park, Sonoma County, California.

Salt crystallization, otherwise known as haloclasty, causes disintegration of rocks when saline solutions seep into cracks and joints in the rocks and evaporate, leaving salt crystals behind. These salt crystals expand as they are heated up, exerting pressure on the confining rock.

Salt crystallization may also take place when solutions decompose rocks (for example, limestone and chalk) to form salt solutions of sodium sulfate or sodium carbonate, of which the moisture evaporates to form their respective salt crystals.

The salts which have proved most effective in disintegrating rocks are sodium sulfate, magnesium sulfate, and calcium chloride. Some of these salts can expand up to three times or even more.

It is normally associated with arid climates where strong heating causes strong evaporation and therefore salt crystallization. It is also common along coasts. An example of salt weathering can be seen in the honeycombed stones in sea wall. Honeycomb is a type of tafoni, a class of cavernous rock weathering structures, which likely develop in large part by chemical and physical salt weathering processes.

Biological Effects on Mechanical Weathering

Living organisms may contribute to mechanical weathering (as well as chemical weathering, see 'biological' weathering below). Lichens and mosses grow on essentially bare rock surfaces and create a more humid chemical microenvironment. The attachment of these organisms to the rock surface enhances physical as well as chemical breakdown of the surface microlayer of the rock. On a larger scale, seedlings sprouting in a crevice and plant roots exert physical pressure as well as providing a pathway for water and chemical infiltration.

Chemical Weathering

Comparison of unweathered (left) and weathered (right) limestone.

Chemical weathering changes the composition of rocks, often transforming them when water interacts with minerals to create various chemical reactions. Chemical weathering is a gradual and ongoing process as the mineralogy of the rock adjusts to the near surface environment. New or *secondary minerals* develop from the original minerals of the rock. In this the processes of oxidation and hydrolysis are most important. Chemical weathering is enhanced by such geological agents as the presence of water and oxygen, as well as by such biological agents as the acids produced by microbial and plant-root metabolism.

The process of mountain block uplift is important in exposing new rock strata to the atmosphere and moisture, enabling important chemical weathering to occur; significant release occurs of Ca^{2+} and other ions into surface waters.

Dissolution and Carbonation

Rainfall is acidic because atmospheric carbon dioxide dissolves in the rainwater producing weak carbonic acid. In unpolluted environments, the rainfall pH is around 5.6. Acid rain occurs when gases such as sulfur dioxide and nitrogen oxides are present in the atmosphere. These oxides react in the rain water to produce stronger acids and can lower the pH to 4.5 or even 3.0. Sulfur dioxide, SO_2, comes from volcanic eruptions or from fossil fuels, can become sulfuric acid within rainwater, which can cause solution weathering to the rocks on which it falls.

A pyrite cube has dissolved away from host rock, leaving gold behind

Some minerals, due to their natural solubility (e.g. evaporites), oxidation potential (iron-rich minerals, such as pyrite), or instability relative to surficial conditions will weather through dissolution naturally, even without acidic water.

One of the most well-known solution weathering processes is carbonation, the process in which atmospheric carbon dioxide leads to solution weathering. Carbonation occurs on rocks which contain calcium carbonate, such as limestone and chalk. This takes place when rain combines with carbon dioxide or an organic acid to form a weak carbonic acid which reacts with calcium carbonate (the limestone) and forms calcium bicarbonate. This process speeds up with a decrease in temperature, not because low temperatures generally drive reactions faster, but because colder water holds more dissolved carbon dioxide gas. Carbonation is therefore a large feature of glacial weathering.

The reactions as follows:

$$CO_2 + H_2O \rightarrow H_2CO_3$$

carbon dioxide + water → carbonic acid

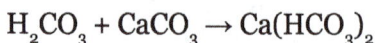

$$H_2CO_3 + CaCO_3 \rightarrow Ca(HCO_3)_2$$

carbonic acid + calcium carbonate → calcium bicarbonate

Carbonation on the surface of well-jointed limestone produces a dissected limestone pavement. This process is most effective along the joints, widening and deepening them.

Hydration

Olivine weathering to iddingsite within a mantle xenolith

Mineral hydration is a form of chemical weathering that involves the rigid attachment of H+ and OH- ions to the atoms and molecules of a mineral.

When rock minerals take up water, the increased volume creates physical stresses within the rock. For example, iron oxides are converted to iron hydroxides and the hydration of anhydrite forms gypsum.

A freshly broken rock shows differential chemical weathering (probably mostly oxidation) progressing inward. This piece of sandstone was found in glacial drift near Angelica, New York

Hydrolysis on Silicates and Carbonates

Hydrolysis is a chemical weathering process affecting silicate and carbonate minerals. In such reactions, pure water ionizes slightly and reacts with silicate minerals. An example reaction:

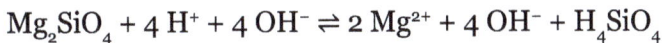

$$Mg_2SiO_4 + 4\ H^+ + 4\ OH^- \rightleftharpoons 2\ Mg^{2+} + 4\ OH^- + H_4SiO_4$$

olivine (forsterite) + four ionized water molecules $\rightleftharpoons$ ions in solution + silicic acid in solution

This reaction theoretically results in complete dissolution of the original mineral, if enough water is available to drive the reaction. In reality, pure water rarely acts as a H^+ donor. Carbon dioxide, though, dissolves readily in water forming a weak acid and H^+ donor.

$$Mg_2SiO_4 + 4\,CO_2 + 4\,H_2O \rightleftharpoons 2\,Mg^{2+} + 4\,HCO_3^- + H_4SiO_4$$

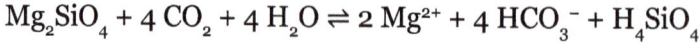

olivine (forsterite) + carbon dioxide + water $\rightleftharpoons$ Magnesium and bicarbonate ions in solution + silicic acid in solution

This hydrolysis reaction is much more common. Carbonic acid is consumed by silicate weathering, resulting in more alkaline solutions because of the bicarbonate. This is an important reaction in controlling the amount of CO_2 in the atmosphere and can affect climate.

Aluminosilicates when subjected to the hydrolysis reaction produce a secondary mineral rather than simply releasing cations.

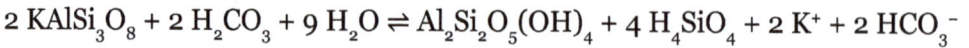

$$2\,KAlSi_3O_8 + 2\,H_2CO_3 + 9\,H_2O \rightleftharpoons Al_2Si_2O_5(OH)_4 + 4\,H_4SiO_4 + 2\,K^+ + 2\,HCO_3^-$$

Orthoclase (aluminosilicate feldspar) + carbonic acid + water $\rightleftharpoons$ Kaolinite (a clay mineral) + silicic acid in solution + potassium and bicarbonate ions in solution

Oxidation

Oxidized pyrite cubes

Within the weathering environment chemical oxidation of a variety of metals occurs. The most commonly observed is the oxidation of Fe^{2+} (iron) and combination with oxygen and water to form Fe^{3+} hydroxides and oxides such as goethite, limonite, and hematite. This gives the affected rocks a reddish-brown coloration on the surface which crumbles easily and weakens the rock. This process is better known as 'rusting', though it is distinct from the rusting of metallic iron. Many other metallic ores and minerals oxidize and hydrate to produce colored deposits, such as chalcopyrites or $CuFeS_2$ oxidizing to copper hydroxide and iron oxides.

Biological Weathering

A number of plants and animals may create chemical weathering through release of acidic compounds, i.e. the effect of moss growing on roofs is classed as weathering.

Mineral weathering can also be initiated and/or accelerated by soil microorganisms. Lichens on rocks are thought to increase chemical weathering rates. For example, an experimental study on hornblende granite in New Jersey, USA, demonstrated a 3x – 4x increase in weathering rate under lichen covered surfaces compared to recently exposed bare rock surfaces.

Biological weathering of basalt by lichen, La Palma.

The most common forms of biological weathering are the release of chelating compounds (i.e. organic acids, siderophores) and of acidifying molecules (i.e. protons, organic acids) by plants so as to break down aluminium and iron containing compounds in the soils beneath them. Decaying remains of dead plants in soil may form organic acids which, when dissolved in water, cause chemical weathering. Extreme release of chelating compounds can easily affect surrounding rocks and soils, and may lead to podsolisation of soils.

The symbiotic mycorrhizal fungi associated with tree root systems can release inorganic nutrients from minerals such as apatite or biotite and transfer these nutrients to the trees, thus contributing to tree nutrition. It was also recently evidenced that bacterial communities can impact mineral stability leading to the release of inorganic nutrients. To date a large range of bacterial strains or communities from diverse genera have been reported to be able to colonize mineral surfaces and/or to weather minerals, and for some of them a plant growth promoting effect was demonstrated. The demonstrated or hypothesised mechanisms used by bacteria to weather minerals include several oxidoreduction and dissolution reactions as well as the production of weathering agents, such as protons, organic acids and chelating molecules.

Building Weathering

Buildings made of any stone, brick or concrete are susceptible to the same weathering agents as any exposed rock surface. Also statues, monuments and ornamental stonework can be badly damaged by natural weathering processes. This is accelerated in areas severely affected by acid rain.

Properties of Well-weathered Soils

Three groups of minerals often remain in well-weathered soils: silicate clays, very resistant end products including iron and aluminium oxide clays, and very resistant primary minerals such as quartz. In highly weathered soils of humid tropical and subtropical regions, the oxides of iron and aluminium, and certain silicate clays with low Si/Al ratios, predominate because most other constituents have been broken down and removed.

References

- Blanco-Canqui, Humberto; Rattan, Lal (2008). "Soil and water conservation". Principles of soil conservation and management. Dordrecht: Springer. pp. 1–20. ISBN 9781402087097.

- Toy, Terrence J.; Foster, George R.; Renard, Kenneth G. (2002). Soil erosion : processes, prediction, measurement, and control. New York, NY: Wiley. ISBN 9780471383697.

- Boardman, John; Poesen, Jean, eds. (2007). Soil Erosion in Europe. Chichester: John Wiley & Sons. ISBN 9780470859117.

- Food and Agriculture Organization (1965). "Types of erosion damage". Soil Erosion by Water: Some Measures for Its Control on Cultivated Lands. United Nations. pp. 23–25. ISBN 978-92-5-100474-6.

- Borah, Deva K. et al. (2008). "Watershed sediment yield". In Garcia, Marcelo H. Sedimentation Engineering: Processes, Measurements, Modeling, and Practice. ASCE Publishing. p. 828. ISBN 978-0-7844-0814-8.

- Nancy D. Gordon (2004-06-01). "Erosion and Scour". Stream hydrology: an introduction for ecologists. ISBN 978-0-470-84357-4.

- Zheng, Xiaojing & Huang, Ning (2009). Mechanics of Wind-Blown Sand Movements. Springer. pp. 7–8. ISBN 978-3-540-88253-4.

- Cornelis, Wim S. (2006). "Hydroclimatology of wind erosion in arid and semi-arid environments". In D'Odorico, Paolo & Porporato, Amilcare. Dryland Ecohydrology. Springer. p. 141. ISBN 978-1-4020-4261-4.

- Blanco-Canqui, Humberto; Rattan, Lal (2008). "Wind erosion". Principles of soil conservation and management. Dordrecht: Springer. pp. 54–80. ISBN 9781402087097.

- Balba, A. Monem (1995). "Desertification: Wind erosion". Management of Problem Soils in Arid Ecosystems. CRC Press. p. 214. ISBN 978-0-87371-811-0.

- Wiggs, Giles F.S. (2011). "Geomorphological hazards in drylands". In Thomas, David S.G. Arid Zone Geomorphology: Process, Form and Change in Drylands. John Wiley & Sons. p. 588. ISBN 978-0-470-71076-0.

- Van Beek, Rens (2008). "Hillside processes: mass wasting, slope stability, and erosion". In Norris, Joanne E. et al. Slope Stability and Erosion Control: Ecotechnological Solutions. Springer. ISBN 978-1-4020-6675-7.

- Gray, Donald H. & Sotir, Robbin B. (1996). "Surficial erosion and mass movement". Biotechnical and Soil Bioengineering Slope Stabilization: A Practical Guide for Erosion Control. John Wiley & Sons. p. 20. ISBN 978-0-471-04978-4.

- Blanco-Canqui, Humberto; Rattan, Lal (2008). "Water erosion". Principles of soil conservation and management. Dordrecht: Springer. pp. 21–53. ISBN 9781402087097.

- Whisenant, Steve G. (2008). "Terrestrial systems". In Perrow Michael R. & Davy, Anthony J. Handbook of Ecological Restoration: Principles of Restoration. Cambridge University Press. p. 89. ISBN 978-0-521-04983-2.

- Wainwright, John & Brazier, Richard E. (2011). "Slope systems". In Thomas, David S.G. Arid Zone Geomorphology: Process, Form and Change in Drylands. John Wiley & Sons. ISBN 978-0-470-71076-0.

- Burbank, Douglas W.; Anderson, Robert S. (2011). "Tectonic and surface uplift rates". Tectonic Geomorphology. John Wiley & Sons. pp. 270–271. ISBN 978-1-4443-4504-9.

- Blanco, Humberto & Lal, Rattan (2010). "Soil and water conservation". Principles of Soil Conservation and Management. Springer. p. 2. ISBN 978-90-481-8529-0.

5

Sedimentary Basin: An Integrated Study

Sedimentary basins are areas filled with sediments. Some of the examples of sedimentary basin are Angola Basin, Los Angeles Basin, Nias Basin and Niger Delta Basin. This chapter helps the readers in developing an in-depth understanding of sedimentary basins.

Sedimentary Basin

Geologic provinces of the World (USGS)

		Oceanic crust:
	Shield	
	Platform	0–20 Ma
	Orogen	20–65 Ma
	Basin	>65 Ma
	Large igneous province	
	Extended crust	

Sedimentary basins are regions of Earth of long-term subsidence creating accommodation space for infilling by sediments. The subsidence can result from a variety of causes that include: the thinning of underlying crust, sedimentary, volcanic, and tectonic loading, and changes in the thickness or density of adjacent lithosphere. Sedimentary basins occur in diverse geological settings usually associated with plate tectonic activity. Basins are classified structurally in various ways, with a primary classifications distinguishing among basins formed in various plate tectonic regime (divergent, convergent, transform, intraplate), the proximity of the basin to the active plate margins, and whether oceanic, continental or transitional crust underlies the basin. Basins formed in different plate tectonic regimes vary in their preservation potential. On oceanic crust, basins are likely to be subducted, while marginal continental basins may be partially preserved, and intracratonic basins have a high probability of preservation. As the sediments are buried, they are subjected to increasing pressure and begin the process of lithification. A number of basins formed in extensional settings can undergo inversion which has accounted for a number of the economically viable oil reserves on earth which were formerly basins.

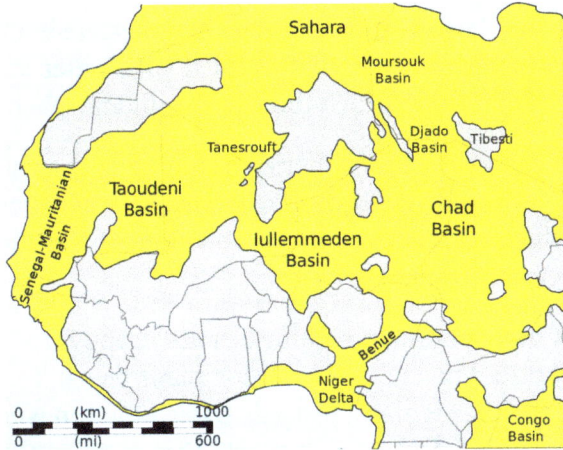

Map of the major sedimentary basins of Central and West Africa.

Methods of Formation

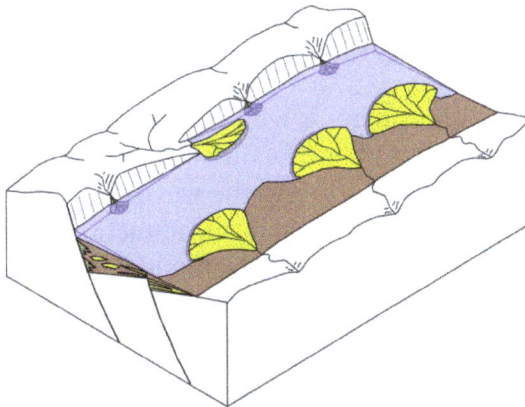

Example of a sedimentary basin in a half-graben.

Sedimentary basins form primarily in convergent, divergent and transform settings. Convergent boundaries create foreland basins through tectonic compression of oceanic and continental crust during lithospheric flexure. Tectonic extension at divergent boundaries where continental rifting is occurring can create a nascent ocean basin leading to either an ocean or the failure of the rift zone. In tectonic strike-slip settings, accommodation spaces occur as transpressional, transtensional or transrotational basins according to the motion of the plates along the fault zone and the local topography pull-apart basins.

Lithospheric Stretching

If the lithosphere is caused to stretch horizontally, by mechanisms such as *ridge-push* or *trench-pull*, the effect is believed to be twofold. The lower, hotter part of the lithosphere will "flow" slowly away from the main area being stretched, whilst the upper, cooler and more brittle crust will tend to fault (crack) and fracture. The combined effect of these two mechanisms is for the Earth's surface in the area of extension to subside, creating a geographical depression which is then often infilled with water and/or sediments. (An analogy might be a piece of rubber, which thins in the middle when stretched.)

An example of a basin caused by lithospheric stretching is the North Sea - also an important location for significant hydrocarbon reserves. Another such feature is the Basin and Range Province which covers most of the USA state of Nevada, forming a series of horst and graben structures.

Another expression of lithospheric stretching results in the formation of ocean basins with central ridges; The Red Sea is in fact an incipient ocean, in a plate tectonic context. The mouth of the Red Sea is also a tectonic triple junction where the Indian Ocean Ridge, Red Sea Rift and East African Rift meet. This is the only place on the planet where such a triple junction in oceanic crust is exposed subaerially. The reason for this is twofold, due to a high thermal buoyancy of the junction, and a local crumpled zone of seafloor crust acting as a dam against the Red Sea.

Lithospheric Compression/Shortening and Flexure

If a load is placed on the lithosphere, it will tend to flex in the manner of an elastic plate. The magnitude of the lithospheric flexure is a function of the imposed load and the *flexural rigidity* of the lithosphere, and the wavelength of flexure is a function of flexural rigidity alone. Flexural rigidity is in itself, a function of the lithospheric mineral composition, thermal regime, and effective elastic thickness. The nature of the load is varied. For instance, the Hawaiian Islands chain of volcanic edifices has sufficient mass to cause deflection in the lithosphere.

The obduction of one tectonic plate onto another also causes a load and often results in the creation of a foreland basin, such as the Po basin next to the Alps in Italy, the Molasse Basin next to the Alps in Germany, or the Ebro basin next to the Pyrenees in Spain.

Strike-slip Deformation

Deformation of the lithosphere in the plane of the earth (i.e. such that faults are vertical) occurs as a result of near horizontal maximum and minimum principal stresses. The resulting zones of subsidence are known as strike-slip or pull-apart basins. Basins formed through strike-slip action occur where a vertical fault plane curves. When the curve in the fault plane moves apart, a region of *transtension* results, creating a basin. Another term for a transtensional basin is a *rhombochasm*. A classic rhombochasm is illustrated by the Dead Sea rift, where northward movement of the Arabian Plate relative to the Anatolian Plate has caused a rhombochasm.

The opposite effect is that of *transpression*, where converging movement of a curved fault plane causes collision of the opposing sides of the fault. An example is the San Bernardino Mountains north of Los Angeles, which result from convergence along a curve in the San Andreas fault system. The Northridge earthquake was caused by vertical movement along local thrust and reverse faults *bunching up* against the bend in the otherwise strike-slip fault environment. In Nigeria, the dominant type of basement rock intersected by wells drilled for hydrocarbons, limestone, or water is granite. The three sedimentary basins in Nigeria are underlain by continental crust except in the Niger delta, where the basement rock is interpreted to be oceanic crust. Most of the wells that penetrated the basement are in the Eastern Dahomey embayment of western Nigeria. A maximum thickness of about 12,000 m of sedimentary rocks is attained in the offshore western Niger delta, but maximum thicknesses of sedimentary rocks are about 2,000 m in the Chad basin and only 500 m in the Sokoto embayment.

Ongoing Development

As more and more sediment is deposited into the basin, the weight of all the newer sediment may cause the basin to subside further because of isostasy. A basin can continue having sediment deposited into it, and continue to subside, for long periods of geological time; this can result in basins many kilometres in thickness. Geologic faults can often occur around the edge of, and within, the basin, as a result of the ongoing slippage and subsidence.

Study of Sedimentary Basins

The study of sedimentary basins as a specific entity in themselves is often referred to as basin modelling or sedimentary basin analysis. The need to understand the processes of basin formation and evolution are not restricted to the purely academic. Indeed, sedimentary basins are the location for almost all of the world's hydrocarbon reserves and as such are the focus of intense commercial interest.

Examples of Sedimentary Basin

Angola Basin

The Angola Basin is located along the West African South Atlantic Margin which extends from Cameroon to Angola. It is characterized as a passive margin that began spreading in the south and then continued upwards throughout the basin. This basin formed during the initial breakup of the supercontinent Pangaea during the early Cretaceous, creating the Atlantic Ocean and causing the formation of the Angola, Cape, and Argentine basins. It is often separated into two units: the Lower Congo Basin, which lies in the northern region and the Kwanza Basin which is in the southern part of the Angola margin. The Angola basin is famous for its "Aptian Salt Basins," a thick layer of evaporites that has influenced topography of the basin since its deposition and acts as an important petroleum reservoir.

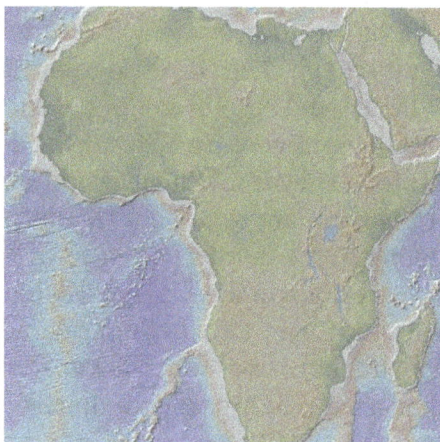

The Angola Basin is located off the west coast of Africa above the Walvis Ridge.

Tectonic Mechanisms

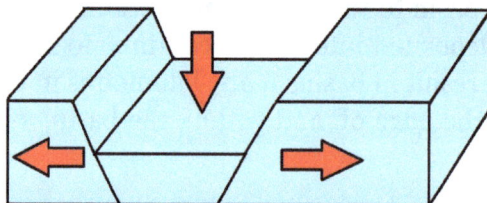

Gravity spreading mechanism based on Peel 2014

Typically divergent boundaries are described as having landward extension, seaward contraction, and translation, however the order of events in this area are difficult to distinguish in such a clear-cut manner. This is due to the fact that areas of the basin are superimposed upon one another, which some interpret to show pulses of deformation and uplift that occur at irregular times and places.

Gravity Spreading

The Angola basin is also highly characterized by gravity spreading where energy is released when the center of gravity lowers as crustal material thins. This spreading mechanism requires at least some deformation as opposed to the breakup of rigid blocks. Gravity spreading is also temporally linked to sediment deposition, so spreading rates should increase during times of high sediment deposition and decrease or halt when there is little to no sediment deposition. As a result any accommodation space created as the margin continues to spread should be filled with sediments.

Salt Tectonics

The evaporite layer present within the basin is responsible for many topographic features that developed since its deposition as salt movement deforms the surrounding bedrock. The driving force behind salt tectonics is thought to be extension governed by gravity. As gravity spreading acts upon the salt layers it causes upslope extension and downslope contraction, which also explains many of the folds and features of the basin. Seismic profiles taken from offshore Angola show many different salt structures such as diapirs, clines, turtle features, and salt walls that show several deformation phases as the salt squeezes upwards when it is deposited upon. Many of the salt forms are associated with early Cretaceous folding and uplift as well as lateral shortening. One of the signature features in the Angola basin are deep troughs that developed as salt dissolved, creating space for sediment fill. The troughs range from the beginning of the Eocene to the end of the Miocene depending on the time of dissolution.

Raft Tectonics

Post-rift deformation is predominantly caused by raft tectonics, a term that is associated with salt detachment when normal fault blocks are widely separated so that the footwall and hanging wall are not in contact, creating large grabens. It is considered one of the most extreme forms of extension, and it highly influenced by gravity spreading and increased sediment loading, major factors which act upon the Angola Basin. In the basin this tectonic mechanism is attributed to three periods of high strain that occurred at approximately 96, 28, and 10 million years ago, and the most recent high strain activity is still ongoing. These high strain rates lasted anywhere from 15 to 36 million years while rafting itself lasted from 7-10 million years.

Geologic History

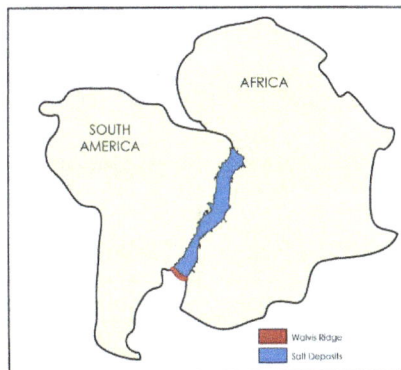

Overview of the rifting in the Cretaceous, approximately 120 million years ago, and placement of salt deposits and the Walvis Ridge adapted from Naafs and Pancost 2014

Mesozoic

The formation of the Angola Basin can be divided into three phases of rifting which took place from approximately 145-113 million years ago from the Jurassic to the Cretaceous. Initial rifting is defined by widespread crustal thinning, normal faulting, and the subsidence of grabens that formed in the upper crust. This was followed by a second rifting phase which was dominated by lithospheric thinning. The final phase of rifting led to the breakup of the lithosphere, initiated seafloor spreading that still acts today, and resulted in the development of oceanic crust.

After the rifting, salt deposited upon the preexisting bedrock. The large amounts of salt in most of the basin make it difficult to determine structures and sedimentary deposits beneath it since seismic does not penetrate through it. Though the salt layer creates some ambiguity most agree that the bedrock is composed of volcanic basalts which are likely a result of rifting or Precambrian crystalline rock. There are two main theories for the environment which called for salt deposition. The first is that the environment was a shallow marine area which after anomalous subsidence events causes rapid salt accumulation. The second hypothesis claims that salt filled a topographic depression much further below sea level. Despite which theory may be correct, it is generally agreed that the basin must have been very restricted from the ocean which allowed the evaporite deposits to be nearly three kilometers thick.

After the salt layer was deposited it was covered by a carbonate layer approximately 112 million years ago. The carbonate formation occurred due to large-scale anoxic events which created organic-rich shales. During this time the basin was hypersaline making it inhospitable for normal marine life, although there may have been a small but stable amount of input of terrestrial fresh water. The source of this freshwater as well as clastic debris was likely from the Kouilou-Niari River which is located in present day Congo. As the rift continued to spread apart Pangaea into the South American and African continents, the Angola Basin opened up further, allowing for better ocean circulation which balanced out the extreme hypersaline conditions to allow for life to develop in the area. Towards the end of the Cretaceous the Congo River began to fill the basin with terrigenous sediments, characterized by many turbidite deposits which replaced most of the carbonate deposits.

Angola Basin Stratigraphy		
Period	Time Interval (Ma)	Sediment Type
Quaternary	15-present	Siltstone/Sandstone
Neogene	34-15	Siltstone/Sandstone
Paleogene	100-34	Shale
Cretaceous	112-100	Carbonates
Cretaceous	117-112	Evaporites

Cenozoic

The Congo River created a much larger impact upon the basin in the Oligocene. The sedimentary fill from the Congo River created a large deep-sea fan where the river enters the ocean, and this fan is still one of the basin's most predominant features. The Oligocene is also characterized by an erosional event that lasted 10-20 million years that is thought to be controlled by upheavals or depressions of crust over a broad area that are caused by mantle convection and hotsopt activity.

From the beginning of the Quaternary to present day much of the sediment is influenced by the Walvis Ridge, a hotspot trail that extends several hundred kilometers off the coast of Africa into the Atlantic Ocean, in addition to the Congo River. During this time the carbonate compensation depth, the depth at which carbonates dissolve, is at a minimum depth of 5400 m, over 1000 meters more than the average depth. This is due to the Walvis Ridge preventing cold Antarctic bottom waters from circulating the basin allowing for the sedimentation of carbonate materials, including microorganisms such as foraminifera and other calcareous microfossils. The Angola Basin is currently well circulated by warm and cold surface currents and undercurrents and is mostly influenced by the Benguela Current, the Equatorial Countercurrents, and the Angola Current.

Subbasins

Lower Congo Basin

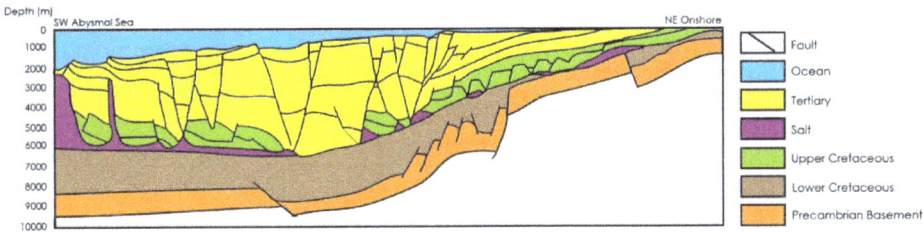

Cross section of the Angola Basin submarine fan from the southwest oceanwards end to the northeart onshore end adapted from Jiang, Wang, and Zheng 2014

The Lower Congo Basin lies in the northern region of the Angola Basin and is largely identified by a sedimentary fan that is fueled by the Congo River and is part of the Ogooue Delta. While the fan is dated to the Oligocene, initial sediment deposition which the fan developed on began in the Cretaceous and contains some of the Aptain salt layer. This fan is one of the largest marine fans in the world as it covers 300,000 square kilometers leading from the mouth of the river into the Atlantic Ocean. Since the fan is mainly composed of turbidite deposits composed for large amounts of sandstone and fine grained muds, it is likely an area that is currently generating hydrocarbons and probably has been for the past 30 million years. This feature is highly dominated by gravity flows where sediment and fluid flow down slope due to gravity.

Kwanza Basin

The Kwanza Basin lies in the lower region of the Angola Basin and can be divided into the inner and outer Kwanza Basins, with the inner basin lying closer to the continent of Africa and the outer basin surrounding the inner basin. Basement structures separate the inner and outer areas of the basin; these structures are named the Flamingo, Ame-tista, and Benguela Platforms which comprises the Atlantic hinge zone. These are areas where the signature salt layer is very thin or absent from the stratigraphic record. The basin's topographic features are mainly affected by salt tectonics, since the salt in most areas was originally over one kilometer thick. There are two main types of salt struc-tures found in the inner Kwanza Basin: narrow salt walls which developed from salt-cored folds, and broad salt walls that formed likely due to major uplift in the area. Many of the salt features dissolved over time which led to the development of sedimentary troughs in the Cenozoic, although fewer troughs did develop as a result of extension.

Hydrocarbons

The basin houses economically important hydrocarbon reservoirs that serve as a source of petroleum. Hydrocarbon generation in the Angola Basin is still an ongoing process that began in the late Cretaceous after the deposition of the thick salt beds. The salt is an important feature in preserving hydrocarbons as it seals in the reservoir and pre-vents it from escaping into the open water. Successful hydrocarbon collection within the Angola margin is associated with pockmarks within the topography that are formed as gas or subsurface water travels upwards through the water column. In December 2000 a research expedition collected gas hydrate specimens from one of the world's largest pockmarks located in the Congo-Angola Basin. The depression was 800 meters in diameter and located 3160 meters below sea level and developed as a result of several smaller pockmarks collapsing into each other. The majority of the hydrocarbons found were gas hydrates composed of 100% methane.

Los Angeles Basin

The Los Angeles Basin is a sedimentary basin located in southern California, in a region known as the Peninsular Ranges. The basin is also connected to an anomalous group of east-west trending chain of basin collectively known as the California Transverse Rang-es. The present basin is a coastal lowland area, whose floor is marked by elongate low ridges and groups of hills that is located on the edge of the Pacific plate. Along with the Los Angeles Basin, the Santa Barbara Channel, the Ventura Basin, the San Fernando Valley, and the San Gabriel Basin lie within the greater southern California region. On the north, northeast, and east, the lowland basin is bound by the Santa Monica Moun-tains mountains and Puente, Elysian, Repetto hills. To the southeast, the basin is bor-dered by the Santa Ana mountains and the San Joaquin Hills. The western boundary of the basin is marked by the Continental Borderland and is part of the onshore portion. The California borderland is characterized by north-west trending offshore ridges and basins. The Los Angeles Basin is notable for its great structural relief and complexity in relation to its geologic youth and small size for its prolific oil production. Yerkes et al. identify 5 major stages of the basin's evolution that begins in the Upper Cretaceous and ends in the Pleistocene. This basin can be classified as an irregular pull-apart basin accompanied by rotational tectonics during the post-early Miocene.

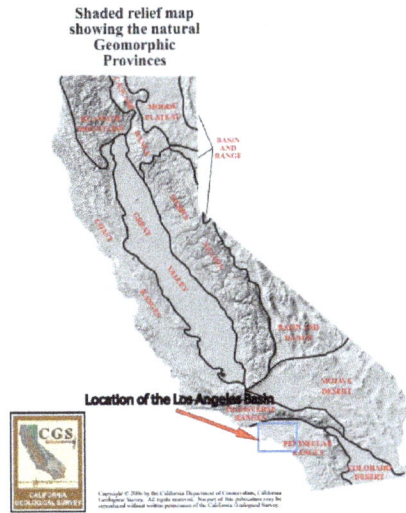

Map of the Los Angeles Basin

Basin Development

Before the formation of the basin, the area that encompasses the Los Angeles basin began above ground. A rapid transgression and regression of the shoreline moved the area to a shallow marine environment. Tectonic instability coupled with volcanic activity in rapidly subsiding areas during the Middle Miocene set the stage for the modern basin. The basin formed in a submarine environment and was later brought back above sea level when the rate of subsidence slowed. There is much discussion in the literature about the geologic time boundaries when each basin forming event took place. While exact ages may not be clear, Yerkes et al. (1965) provided a general timeline to categorize the sequence of depositional events in the LA Basin's evolution and they are as follows:

Phase 1: Pre-extension

During pre-Turonian, metamorphosed sedimentary and volcanic rocks are present that serve as the two major basement rock units for the LA Basin. Large-scale movement along the Newport-Inglewood zone juxtaposed the two bedrock units along the east and west margins. During this phase, the basin was above sea level.

Phase 2: Pre-basin Phase of Deposition

The hallmarks of this phase were successive shoreline transgression and regression cycles. Deposition of older marine and non-marine sediments began to fill the basin. Towards the end of this phase, the shoreline began to retreat and deposition continued.

Phase 3: Basin Inception

After the deposition of the pre-Turonian units, there was a large emergence and erosion that can be observed as a major unconformity at the base of the middle Miocene units. Emergence did not occur at the same rate or in all sections of the basin. During this time, the basin was covered by a marine embayment. Rivers sourced in the highlands brought large amounts of detritus to the northeastern edge of the basin. During this period, the Topanga formation was also being deposited.

Phase 4: Principal Phase of Subsidence and Deposition

The present form and structural relief of the basin was largely established during this phase of accelerated subsidence and deposition which occurred during the late Miocene and continued through the early Pleistocene. Clastic sedimentary rocks from the highland areas (to the north and east) moved down the submarine slopes and infilled the basin floor. Subsidence and sedimentation most likely began in the southern portion basin. Subsidence and Deposition occurred simultaneously, without interruption, until the late Pliocene. Until the rate of deposition gradually overtook the rate of subsidence, and the sea level began to fall. Towards the end of this phase, the margins of the basin began to rise above sea level. During the early Pleistocene, deposition began to outpace subsidence in the depressed parts of the basin and the shoreline began to move southward. This phase also had movement along the Newport-Inglewood fault zone that resulted in the initiation of the modern basin. This movement caused the southwestern block to be uplifted relative to the central basin block.

Phase 5: Basin Disruption

The central part of the basin continued to experience sediment deposition through the Pleistocene from flooding and erosional debris from the surrounding mountains and Puente Hills. This infill was responsible for the final retreat of the shoreline from the basin. Deposition in the Holocene is characterized by non marine gravel, sand and silt. This phase also includes the late stage compressional deformation responsible for the formation of the hydrocarbon traps.

Basin Blocks

Four major faults are present in the region and divide the basin in the central, northwest, southwest, and northeast structural blocks. These blocks not only denote their geographic location, but they indicate the strata present and major structural features. The southwestern block was uplifted prior to the middle Miocene and is composed mostly of marine strata and contains two major anticlines. This block also contains the steeply-dipping Palos Verdes Hills fault zone. The middle Miocene volcanics can be seen locally within the southwest block. The northwestern block consists of clastic marine sediments of Late Cretaceous to Pleistocene age. Middle Miocene volcanics are also present. This block has a broad anticline that is truncated by the Santa Monica fault zone. The central block contains both marine and non-marine clastic rock units interbedded with volcanic rocks that are late Cretaceous to Pliocene in age. Pliocene and Quaternary strata are most visible within the central block. Structurally, there is a synclinal trough. The northeastern block contains fine to coarse grained clastic marine rocks of Cenozoic age. Locally, middle Miocene volcanics can be seen as well as Eocene to Miocene aged non-marine sedimentary rocks. There is also an anticline in the northeastern block.

Basin Stratigraphy

Los Angeles Basin Stratigraphy

Epoch	Formation	Formation Members
Holocene	Alluvium	
Pleistocene	San Pedro/ Sagus Formations	
Pliocene	Fernando	Pico Formation Repetto Formation
Upper Miocene	Monterey/ Modelo Formations	
Middle Miocene	Puente	Sycamore Canyon Member Yorba Member Soquel Member La Vida Member
Lower Miocene	Topanga	Topanga Canyon Formation Conejo Volcanics Calabasas Formation
Oligocene	Vaqueros/ Sespe Formations	

Table 1: This table shows the major stratigraphic units that can be found within the Los Angeles Basin.

Cenozoic Basin Stratigraphy

Homogeneous evolution of this basin did not occur due to dynamic tectonic activity. Despite the active setting, there are over 9,100 m of strata within the basin. The dynamic setting was also responsible for the heterogeneous deposition of each formation. It is common for rock units of the same depositional event to have different names in different locations within the basin. This may be a result of large variation in clast size as with the upper Pliocene Pico Formation in the northwestern part of the basin and the Upper Fernando Formation in the southwest part of the basin. The Los Angeles Basin contains what is known as the "Great Unconformity" which has been interpreted as a large scale erosional event in the basement rock unit. This unconformity is used to correlate strata throughout the basin. The record of the Cenozoic activity begins above this unconformity. The stratigraphic record for this basin indicates that it began as a non-marine environment and then transgressed to a deep ocean system. The oldest basement units of this basin are of both sedimentary and igneous origin. The sedimentary unit was metamorphosed as a result of slippage of the Newport-Inglewood fault and is known as the Catalina Schist. The Catalina Schist can be found on the southwestern edge of the basin and is predominantly a chlorite-quartz schist. Closer to the Newport-Inglewood fault zone, garnet-bearing schists and metagabbros occur. The Santa Monica Slate can be observed in the northwestern block of the basin. The eastern complex is characterized by Santiago Peak Volcanics. This rock unit contains andesitic breccias, flow, agglomerates and tuffs.

The Sespe Formation is the first to appear above the "Great Unconformity" and is marked by interbedded mudstones, sandstones and pebbly sandstones. This bed sequence indicates an alluvial fan, meandering stream or braided stream origin. Upward from the Sespe Formation toward the Vaqueros, the grains become finer and the beds become thinner; indicating a transition to a shallow marine environment. The Vaqueros Formation is marked by two sandstone, siltstone and shale units. There are also characteristic mollusk fossils that indicate the area was dominately shallow marine.

The Topanga Group is the next major formation in the stratigraphic sequence and infills the topography on older rocks. It is a mixed sedimentary and volcanic unit whose base is an erosional unconformity. The unit consists of 3 parts: First is a basal marine conglomeratic sandstone, followed by a dominantly basaltic middle layer of multiple submarine lava flows and tuffs. The youngest part of this unit is a sedimentary breccia, conglomerate, sandstone, and a siltstone. The earliest deposits of the Topanga Group appear to reflect the continuation of a shift in shoreline that can be seen in both the Sespe and Vaqueros formations. Eruptions from one or more of volcanic centers locally and temporarily interrupted sedimentation.

The Puente Formation is a deep-marine formation that is characterized by pro-delta sediments and an overlapping fan system. This unit lies above the Topanga Group giving it a Late Miocene depositional age and is divided into four members. The La Vida Member is a micaceous, platy siltstone with subordinate amounts of thin-bedded feldspathic sandstone. The next member is the Soquel, which is a thick bedded to massive micaceous sandstone. Locally abundant siltstone, conglomerate, and intraformational breccia can also be seen in this member. Above the Soquel lies the Yorba Member. This member is a sandy siltstone that is interbedded with a fine-grained sandstone. The Sycamore Canyon Member contains lenses of conglomerate, conglomeratic sandstone, and sandstone. Sandy siltstone and fine-grained sandstones are interbedded with the aforementioned rock types.

The Monterey Formation is characterized by abnormally high silica content compared to most clastic rocks. There are also silica-cemented rocks known as porcelanite and porcelanite shale. While this formation has distinguishable beds, there are many shale, sandstone, and mudstone beds that have normal amounts of silica. This sequence of this formation indicates an off-shore marine environment.

The Fernando Formation is split into two sub-facies known as the Pico and Repetto Members. These members represent a distinct change in the depositional environment and are of Pleistocene age. The Repetto is the older of the two members and is composed of interbedded fine to coarse grained siltstone, mudstone, and sandstone. The Pico Member is mostly made of massive siltstones and sandstones interbedded with minor silty-sandstones. Holocene Alluvium and Quaternary sediments is a largely unconsolidated unit and is composed mostly of gravel and floodplain sediments. The sediments that mark the top of the basin can be found in modern streams/rivers and at the base of the foothills.

Tectonic Setting

The history of this basin begins with the subduction of Pacific plate underneath the North American plate in the beginning of the Mesozoic. During this subduction event, two smaller plates, the Monterey and Juan de Fuca plates, also began to subduct underneath the North American plate. Around 20Ma, the Monterey plate attached to and followed the motion of the Pacific plate. Later, subduction of the Pacific-Monterey ceased and the plate margin was converted to a transform boundary. The North America/Pacific-Monterey transform boundary began to move north and created crustal extension. This rifting was accompanied with the rotation of the western Transverse Ranges. This rotation is responsible for the placement and northwest-southeast orientation of the LA Basin. Early in the Miocene, before deposition of the Topanga, high heat flow and transtension caused the extension of the basin. As the crust thinned, the basin began to subside from isostatic pressure as a result of large amounts of sediment deposition.

Because the basin lies on the boundary of the Transverse and Peninsular Ranges, this basin experiences both compressional and strike slip tectonics. During the early Pliocene, also identified as the "Basin Disruption" phase, deformation and folding occurred as a result of fault movement and a slight rotation event. While movement along the San Andreas Fault is responsible for the placement of the basin, it is the Whittier and Newport-Inglewood faults that have dictated the seismic behavior within the basin.

Earthquakes

The Los Angeles basin is still active tectonically and the region continues to experience earthquakes as a result. Due to the number of faults and fault splays, seismic activity is not concentrated in one particular area. The cities that are overlain by the Newport-Inglewood and Whittier fault zones have a higher probability of experiencing seismic activity. The region experiences earthquakes that are mostly mild (magnitude ≤2.25). However moderate earthquakes (magnitude 4.9 to 6.4) have been reported. Earthquakes of moderate magnitude are very infrequent.

Basin Features

Structural Features of the Los Angeles Basin

The Newport–Inglewood Fault Zone

This fault zone is the most notable feature within the basin that is a single strand with local (fault) splays. The fault zone is also marked by low hills, scarps, and ten anticlinal folds in a right-stepping en echelon pattern. It is located in the southwest portion of the basin and is a strike-slip margin. There are several oil fields that run parallel to this fault.

The Whittier Fault

This fault lies on the eastern border of the basin and mergers with the Elsinore Fault in the canyon of the Santa Ana river. This fault is a reverse right-oblique fault. It is most known for the Whittier, Brea-Olinda, Sansinena, oil fields. There is an anticline that runs parallel to the Whittier fault that is evidence for compressional deformation during the late Miocene to early Pliocene. Thinning and pinch-out of the Pliocene sandstones are evidence for uplift during this same time period.

The Anaheim Nose

The Anaheim nose is a subsurface feature that was discovered by geophysical surveys and exploratory drilling in 1930. It is a mid-Miocene fault block that revealed a northwest trending ridge of Paleocene age rocks. This structural feature is important because it revealed oil traps and orientation of the beds indicate the age of subsidence in this portion of the basin.

The Wilmington Anticline

This particular anticline is the most notable subsurface feature within the basin. Deformation events such as erosion of the uplifted crustal blocks, initiation of various faults, and the development of the submarine channel led to the anticline's formation. Fold initiation began in the late-Miocene to early Pliocene period of deformation. There are many other anticlines within the basin and isopach data suggests that the formation of these folds occurred mostly during the Pliocene.

The La Brea Tar Pits

The La Brea Tar Pits are pools of stagnant asphaltum that have been found on the basin's surface. These "pools" are important because hundreds of thousands of late Pleistocene bones and plants have been found. These pits allowed scientists to better understand the ecosystem at that particular point in the geologic past.

Petroleum

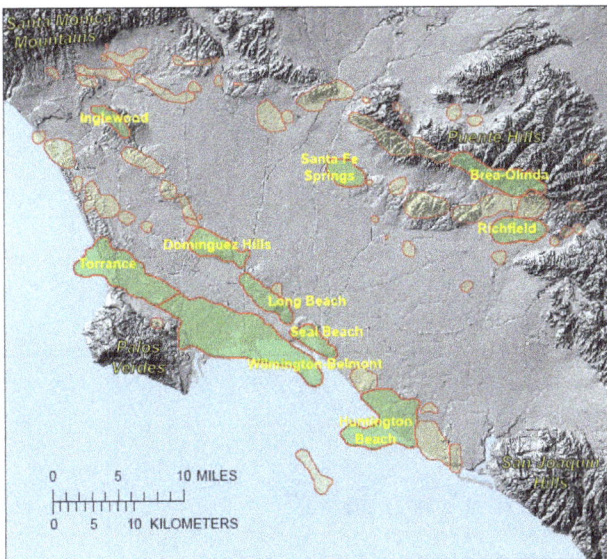

LA Basin Oil Fields, USGS

Accumulations of oil and gas occur almost wholly within strata of the younger sequence and in areas that are within or adjacent to the coastal belt. The Puente formation has proved to be the most notable reservoir for petroleum in the basin. The primary reason for the high abundance of oil is because the oil sands are well saturated within the basin. The thickness of these oil sands range from hundreds to thousands of feet. Anticlines and faulted anticlines are the structural features that are also responsible for trapping oil. The first reported oil-producing well was discovered in 1892 on the land that is presently beneath Dodgers Stadium. This basin was responsible for half of the states oil production until the (90's?). This is remarkable due to the relatively small size and youth of the basin. The basin currently has about 40 active oil fields that collectively have 4,000 operating wells. In 1904, there were over 1,150 wells in the city of Los Angeles alone. Tight spacing and continued pumping of the wells resulted in most of the wells to dry up. Most recent data indicates that 255 million barrels of oil were produced in 2013. This is a large decline from the almost 1 billion barrels per year produced in the late 1970s.

Los Angeles City Oil Field, 1905

Nias Basin

Nias Basin

The Nias Basin (also known as the West Sumatra or Sibolga Basin) is a Forearc basin located off the western coast of Sumatra, Indonesia, in the Indian Ocean. The name is derived from the island that bounds its western edge, the island of Nias. The Nias Basin, the island of Nias (which is a subaerial part of the accretionary complex), and the offshore, submarine accretionary complex, together form a Forearc region on the Sunda Plate/Indo-Australian Plate collisional/subduction boundary. The Forearc region is the area between an oceanic trench and its associated volcanic arc. The oceanic trench associated with the Nias Basin is the Sunda Trench, and the associated volcanic arc is the Sunda Arc.

The Nias Basin itself is structurally bounded to the west by the Mentawai Fault and bounded to the east by the Volcanic Arc island of Sumatra. It is a geologically independent basin from its neighbor basins; the Simeulue Basin to the north, and the Mentawai and Enggano Basin to the south.The Nias Basin spans ~250 kilometer length-wise, and ~100 kilometers width-wise. Overall, the Nias Basin can be divided into two sub-basins; the Singkel Basin to the north, and the Pini Basin to the south. These basins are distinguished by their independent development during the early formation of the primary basin, but later consolidated when subsidence of the area was more unified over the whole Nias Basin region.

Basin Formation

The history of the Nias Basin begins with the initial subduction of the Indo-Australian plate underneath the Sunda Plate. Subduction of this plate, which was rich in water and volatiles, caused flux melting to occur in the mantle. This new magma eventually rose through the overriding plate, and formed the Sunda Arc. The sediment supply, and convergence rate in this case were adequate enough to allow for the Forearc to develop into an Accretionary Convergent Margin. This type of Forearc results in an accretionary complex forming along the Forearc-trench boundary, which resulted in uplift of that region. Uplift due to accretion formed a Forearc Ridge, which the island of Nias is part of. The resulting depression that formed between the Forearc Ridge and the primary volcanic ridge (Sunda Arc), allows for sediment deposition in that region. This depression in this case formed the Nias Basin. Presently, the Nias Basin lies under about 610 meters of water at its deepest point.

The Nias Basin can be divided into 2 sub-basins; the Pini Basin to the South, and the Singkel Basin to the North. These basins evolved independently from each other originally during a time of sea-level regression. These sub-basins themselves are bounded by normal faults, which were formed due to depression of the region during sediment subsidence, and extension events.

The Nias Basin itself is relatively shallow compared to its surrounding basins. This could be due to more carbonate reef activity in this region creating thicker carbonate deposits in the Nias Basin, resulting in more sediment deposition, and shallower water depth.

Stratigraphy

The basement rock of the Nias Basin, at its greatest depth lies at about 4-6 kilometers under the seafloor. This basement rock was determined to be the remains of an older accretionary complex that formed before the Indo-Australian/Sunda collision. The stratigraphic sequence of this basin can be divided into 3 primary sequences.

1st sequence - Pre-Neogene

The lowest and oldest sequence consists of rocks that date to around the late Eocene. These rocks include pyritic shales, dolomitic limestones, and calcareous mudstones. Above these rocks lie various volcanoclastic sandstones and claystones from the early Oligocene.

Depth	Lithology	Age	Age
100m	Limestone	Pliocene-Recent	
200m	Limestone	Pliocene-Recent	
300m	Limestone	Pliocene-Recent	
400m	Limestone	Pliocene-Recent	
500m	Limestone	Pliocene-Recent	
600m	Limestone	Pliocene-Recent	
700m	Limestone	Pliocene-Recent	
800m	Limestone	Pliocene-Recent	
900m	Limestone	Pliocene-Recent	
1000m	Mudstone	Mid-Pliocene	
1100m	Sandstone	Mid-Pliocene	Unconformity
1200m	Limestone	Lower Miocene	
1300m	Limestone	Lower Miocene	
1400m	Limestone	Lower Miocene	
1500m	Limestone	Lower Miocene	
1600m	Limestone	Lower Miocene	
1700m	Limestone	Lower Miocene	
1800m	Limestone	Lower Miocene	
1900m	Limestone	Lower Miocene	
2000m	Sandstone	Lower Miocene	
2100m	Sandstone	Lower Miocene	
2200m	Shale	Oligocene	Unconformity
2300m	Limestone	Late Eocene	

Stratigraphy of Nias Basin

2nd Sequence - Lower Miocene-Upper Miocene

Above the Pre-Neogene sequence lies a major unconformity, overlain by early Miocene rocks. This unconformity is the result of the area being subjected to subaerial erosion, but then was followed by a marine transgression, which deposited the overlaying Miocene rocks. This transgression originally deposited near-shore sands, followed by shallow water siltstones. Towards the middle Miocene, the region developed into a carbonate shelf, and these carbonates make up the rest of this sequence. During the late Miocene, these carbonates were buried by large quantities of clastic sediments originating from the uplift of Sumatra during this time. These sediments deposited faster than the rate of subsidence in the region, creating a continental shelf and slope in the west of the basin. The newly formed continental slope deposited turbidites over the early-middle Miocene sediment that the shelf did not cover. This Sequence is about 1000 meters thick. After the middle Miocene, there lies an unconformity which represents a ~10 million year depositional hiatus.

3rd Sequence - Lower Pliocene-Recent

This sequence, which is also ~1000 meters thick begins with an unconformity which lies at the top of the previous sequence. During this time, clastic sediments continue to be deposited, resulting in the shelf prograding further west, while the deep regions of the basin see continued turbidite deposition.

Stratigraphic Interpretations

Beginning with the 1st sequence, the lithologies are consistent with the formation of the basin along with the volcanic arc, the deposition of the volcanic arc sediments, and ultimate uplift of the basin region, which causes the erosional unconformity. The 2nd sequence stratigraphy displays a marine transgression, and ends with an erosional unconformity. This unconformity could be caused by subaerial erosion due to a regression following the transgression, but it is unknown because that stratigraphy has been eroded. The 3rd sequence shows a marine transgressive event following the previous unconformity, leading into the present day.

Geologic Structures

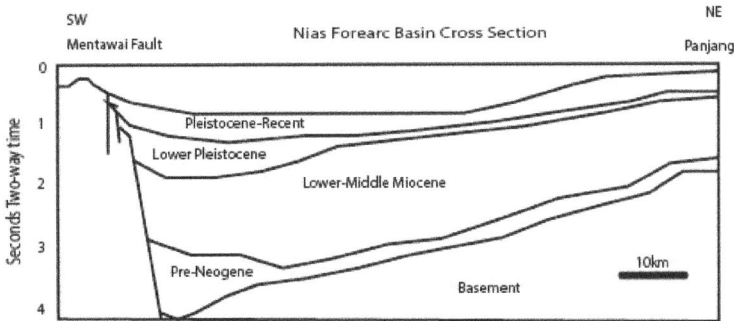

Cross Section Of Nias Basin

The Nias Basin is an asymmetric basin, with the Mentawai Fault bounding its Western edge. The Mentawai Fault is a transform fault that parallels the Sunda Trench, and separates the accretionary complex of the fore-arc, from the fore-arc basin. This fault is the result of the oblique subduction nature of the Indo-Australian Plate. The Mentawai Fault also has a normal fault component to its behavior as well, assisting in the formation of the fore-arc basin, as well as creating more space for sediment deposition and subsidence. The motion of the Mentawai Fault also results in accessory faults within the Nias Basin itself. These faults are also normal/transform in nature, and are the faults that bound the 2 sub-basins within the Nias Basin. These smaller faults form ~30 degrees to the Mentawai Fault, and create graben-like structures.

Natural Resources

The forearc basin depositional environment is characterized by a low geo-thermal gradient. This relatively cool environment, along with the shallow nature of the sea-floor as well as the depositional beds provide a welcoming setting for the creation of hydrocarbons. The coal beds above the second unconformity would be the source of these hydrocarbons, but this area has only just begun being explored for oil/natural gas.

Niger Delta Basin (Geology)

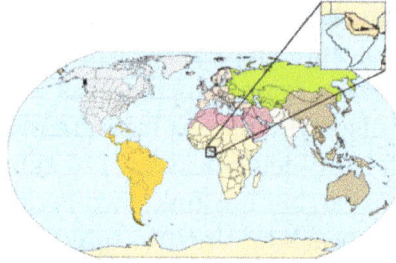

The Niger Delta Basin is located in the Gulf of Guinea on the west coast of Africa.

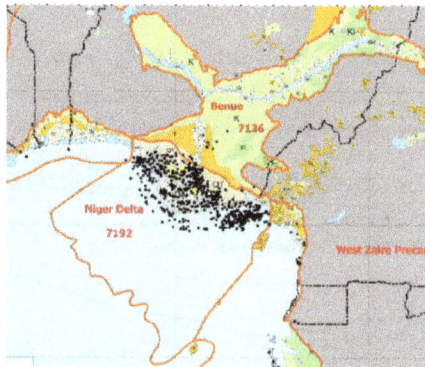

Geologic map of the Niger Delta Basin and the Benue trough, and the oil fields in the region.

The Niger Delta Basin, also referred to as the Niger Delta province, is an extensional rift basin located in the Niger Delta and the Gulf of Guinea on the passive continental margin near the western coast of Nigeria with suspected or proven access to Cameroon, Equatorial Guinea and São Tomé and Príncipe. This basin is very complex, and it carries high economic value as it contains a very productive petroleum system. The Niger delta basin is one of the largest subaerial basins in Africa. It has a subaerial area of about 75,000 km², a total area of 300,000 km², and a sediment fill of 500,000 km³. The sediment fill has a depth between 9–12 km. It is composed of several different geologic formations that indicate how this basin could have formed, as well as the regional and large scale tectonics of the area. The Niger Delta Basin is an extensional basin surrounded by many other basins in the area that all formed from similar processes. The Niger Delta Basin lies in the south westernmost part of a larger tectonic structure, the Benue Trough. The other side of the basin is bounded by the Cameroon Volcanic Line and the transform passive continental margin.

Basin Formation

The Niger Delta Basin was formed by a failed rift junction during separation of the South American plate and the African plate, as well as the opening of the South Atlantic. Rifting in this basin started in the late Jurassic and ended in the mid Cretaceous. As rifting continued, several faults formed, many of them thrust faults. Also at this time we have the deposition of the syn-rift sands and then shales in the late cretaceous. This shows that there was a regression in the early basin. By this time the basin has been undergoing extension by high angle normal faults and fault block rotation. Then at beginning of the Paleocene there was a large transgression. Then in the Paleocene the Akata formation was deposited. In the Eocene the Agbada formation was deposited. This caused the underlying shale Akata Formation to be squeezed into shale diapirs. Then in the Oligocene the Benin formation was deposited and it still being deposited today. The overall basin is divided into a few different zones due to its tectonic structure. There is an extensional zone, which lies on the continental shelf, that is caused by the thickened crust. There is a transition zone, and then there is a contraction zone, which lies in the deep sea part of the basin.

Lithology

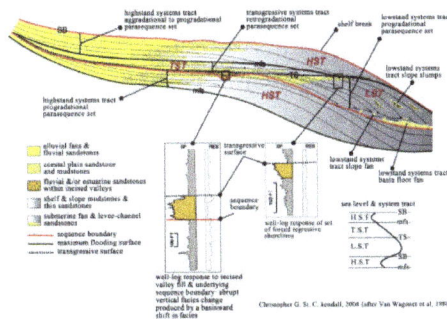

Sealevel highstand and lowstand mapped on a crossection of the basin

The sediment fill in the Niger Delta basin is characterized by three major depobelts. These three cycles show that the basin experienced an overall regression throughout time as the sediments go from deep sea mud sized grains to fluvial denser sand sized grains. The lithologies of the area experience changes due to several factors. One factor would be the types of sediment coming through the delta, which could be influenced by sea level, or maybe volcanic activity in the area. The type of environment of deposition will also change the sediment type. The early Cretaceous sediments were thought to be from a tide dominated system that were deposited on a concave shoreline, and throughout time the shoreline has become convexed and it is currently a wave dominated system.

Basement

The oceanic basement rock is the oldest rock in the basin and is basaltic in composition. It is of the pre-rift time period. Also closer to the coast you have precambrian continental basement.

Cretaceous

There is a section of rock in this basin from the middle to late cretaceous that there is not much information on due to extreme depth. It is believed to be composed of sediments from a tide dominated coastline, and there are believed to be several layers of shales, although there distribution is not known.

Akata Formation

The Akata Formation is Paleocene in age. It is composed of thick shales, turbidite sands, and small amounts of silt and clay. It is the mobile formation that is squeezed into shale diapirs in the basin that are formed form being over pressured and not being dehydrated properly. The Akata formation formed during lowstands in sealevel and in oxygen deficient conditions. This formation is estimated to be up to 7000 meters thick.

Agbada Formation

The Agbada Formation dates back to Eocene in age. It is a marine facies defined by both freshwater and deep sea characteristics. This is the major oil and natural gas bearing facies in the basin. The hydrocarbons in this layer formed when this layer of rock became subaerial and was covered in a swamp type of environment that contained lots of organics. It is estimated to be 3700 meters thick.

Benin Formation

The Benin Formation is Oligocene and younger in age. It is composed of continental flood plain sands and alluvial deposits. It is estimated to be up to 2000 meters thick.

Tectonic Structures

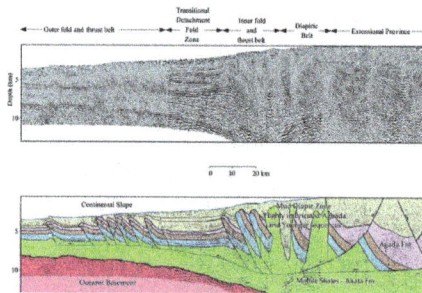

Tectonic Structures drawn over a seismic profile of the Niger Delta Basin

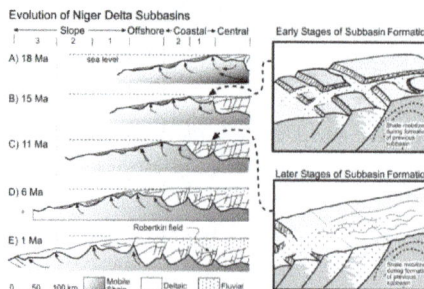

Evolution of Niger Delta subbasins as mobile shales migrate towards the continental slope.

The tectonic structures in the Niger Delta Basin are very typical of an extensional rift system, but the added shale diapirism due to compression makes this basin different. The main method of deformation is due to gravitational collapse of the basin, although the older faulting and deformation in the basin are related to the continental breakup and rifting of the African plate and South American plates. The overall basin is divided into a few different zones due to its tectonic structure. There is an extensional zone, which lies on the continental shelf, that is caused by the thickened crust. There is a transition zone, and then there is a contraction zone, which lies in the deep sea part of the basin.

Basin Inversion

Basin Inversion is caused by uplift and/or compression in this basin. The compression is caused by the toe detachment of the shale diapirs. Basin inversion forms anticline structures, which serve as a great oil trap. Clay smears in the sediments seal the formations so oil does not escape out.

Basinward dipping reflectors

Basinward dipping reflectors are a common feature of extensional type rift basins. As fault blocks extend they rotate to dip towards the center of the basin. At the top of these fault blocks sub basins can form.

Shale Diapirs

The shale diapirs are from the Akata formation. This structure is formed due to the improper dehydration of the formation and the over pressuring by the overlying and denser Agbada formation.

High angle normal faulting

High angle normal faulting is a feature of the extensional portion of the rifting in this basin. It is considered a growth fault and the feature lies closer towards the basins edge and transitions to the toe detachment faulting as you continue down the basin.

Petroleum

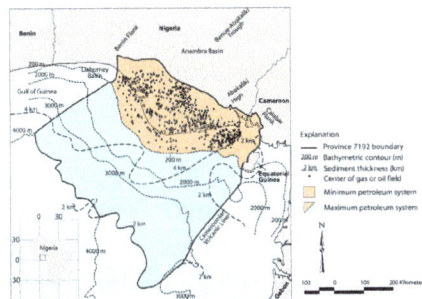

Figure 1 Index map of Nigeria and Cameroon. Map of the Niger Delta showing Province outline (maximum petroleum system); bounding structural features; minimum petroleum system as defined by oil and gas field center points (data from Petroconsultants, 1996a), 200, 2000, 3000, and 4000 m bathymetric contours; and 2 and 4 km sediment thickness.

Minimum and maximum petroleum system

The Niger Delta Basin produces around 2 million barrels of oil per day. The entire system is predicted to contain 34.5 billion barrels of oil and 94 trillion feet^3 of natural gas. This area is still very heavily explored by oil companies today. It is one of the larges oil producers in the world.

Basin Modelling

Basin modelling is the term broadly applied to a group of geological disciplines that can be used to analyse the formation and evolution of sedimentary basins, often but not exclusively to aid evaluation of potential hydrocarbon reserves.

At its most basic, a basin modelling exercise must assess:

1. The burial history of the basin.

2. The thermal history of the basin.

3. The maturity history of the source rocks.

4. The expulsion, migration and trapping of hydrocarbons.

By doing so, valuable inferences can be made about such matters as hydrocarbon generation and timing, maturity of potential source rocks and migration paths of expelled hydrocarbons.

Sedimentary Basin Analysis

Sedimentary basin analysis is a geologic method by which the history of a sedimentary basin is revealed, by analyzing the sediment fill itself. Aspects of the sediment, namely its composition, primary structures, and internal architecture, can be synthesized into a history of the basin fill. Such a synthesis can reveal how the basin formed, how the sediment fill was transported or precipitated, and reveal sources of the sediment fill. From such syntheses models can be developed to explain broad basin formation mechanisms. Examples of such basinal environments include backarc, forearc, passive margin, epicontinental, and extensional basins.

Sedimentary basin analysis is largely conducted by two types of geologists who have slightly different goals and approaches. The petroleum geologist, whose ultimate goal is to determine the possible presence and extent of hydrocarbons and hydrocarbon-bearing rocks in a basin, and the academic geologist, who may be concerned with any or all facets of a basin's evolution. Petroleum industry basin analysis is often conducted on subterranean basins through the use of reflection seismology and data from well logging. Academic geologists study subterranean basins as well as those basins which have been exhumed and dissected by subsequent tectonic events. Thus academics sometimes use petroleum industry techniques, but in many cases they are able to study rocks at the surface. Techniques used to study surficial sedimentary rocks include: measuring stratigraphic sections, identifying sedimentary depositional environments and constructing a geologic map.

An important tool in sedimentary basin analysis is sequence stratigraphy, in which various sedimentary sequences are related to pervasive changes in sea level and sediment supply.

References

- Allen, Philip A.; John R. Allen (2008). Basin analysis : principles and applications (2. ed., [Nachdr.] ed.). Malden, MA [u.a.]: Blackwell. ISBN 978-0-6320-5207-3.

- Cathy J. Busby and Raymond V. Ingersoll, ed. (1995). Tectonics of sedimentary basins. Cambridge, Mass. [u.a.]: Blackwell Science. ISBN 978-0865422452.

- Duppenbecker S. J. and Eliffe J. E., Basin Modelling: Practice and Progress, Geological Society Special Publication, (1998). ISBN 1-86239-008-8

- Hantschel, T. and Kauerauf, A.I., Fundamentals of Basin and Petroleum Systems Modeling, Springer (2009). ISBN 978-3-540-72317-2

- Duppenbecker S. J. and Eliffe J. E., Basin Modelling: Practice and Progress, Geological Society Special Publication, (1998). ISBN 1-86239-008-8

- "Walvis Ridge MV1203 Expedtion: Understanding 130 Million Years of Hotspot Volcanism in the SE Atlantic". earthref.org. National Science Foundation. Retrieved 22 February 2015.

- "Petroleum & Other Liquids: California- Los Angeles Basin Onshore Crude Oil Provided Reserves". U.S. Energy Information Administration. U.S. Department of Energy. Retrieved 18 February 2015.

Stratigraphy: An Integrated Study

The chapter deals with the major principles of stratigraphy. Some of these principles are law of superposition, principle of original horizontality and cross-cutting relationships. The law of superposition is a principle that forms the bases of geology, archaeology and some other fields also. The principle of original horizontality states that layers of sediments are deposited horizontally because of gravity. The aspects explained in this section are of vital importance and provides a better understanding of sedimentology.

Stratigraphy

Stratigraphy is a branch of geology which studies rock layers (strata) and layering (stratification). It is primarily used in the study of sedimentary and layered volcanic rocks. Stratigraphy has two related subfields: lithologic stratigraphy or lithostratigraphy, and biologic stratigraphy or biostratigraphy.

Law of Superposition

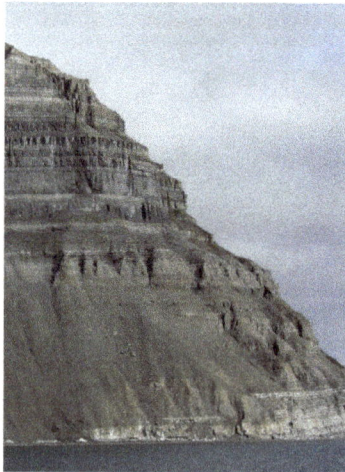

Layer upon layer of rocks on north shore of Isfjord, Svalbard, Norway. Since there is no overturning, the rock at the bottom is older than the rock on the top by the Law of Superposition.

The law of superposition is an axiom that forms one of the bases of the sciences of geology, archaeology, and other fields dealing with geological stratigraphy. In its plainest form, it states that in undeformed stratigraphic sequences, the oldest strata will be at the bottom of the sequence. This is important to stratigraphic dating, which assumes that the law of superposition holds true and that an object cannot be older than the materials of which it is composed. The law was first proposed in the 17th century by the Danish scientist Nicolas Steno.

Archaeological Considerations

Superposition in archaeology and especially in stratification use during excavation is slightly different as the processes involved in laying down archaeological strata are somewhat different from geological processes. Man made intrusions and activity in the archaeological record need not form chronologically from top to bottom or be deformed from the horizontal as natural strata are by equivalent processes. Some archaeological strata (often termed as contexts or layers) are created by undercutting previous strata. An example would be that the silt back-fill of an underground drain would form some time after the ground immediately above it. Other examples of non vertical superposition would be modifications to standing structures such as the creation of new doors and windows in a wall. Superposition in archaeology requires a degree of interpretation to correctly identify chronological sequences and in this sense superposition in archaeology is more dynamic and multi-dimensional.

Principle of Original Horizontality

A stratigraphic section of Ordovician rock exposed in central Tennessee, USA. The sediments composing these rocks were formed in an ocean and deposited in horizontal layers.

The Permian through Jurassic stratigraphy of the Colorado Plateau area of southeastern Utah is a great example of Original Horizontality. These strata make up much of the famous prominent rock formations in widely spaced protected areas such as Capitol

Reef National Park and Canyonlands National Park. From top to bottom: Rounded tan domes of the Navajo Sandstone, layered red Kayenta Formation, cliff-forming, vertically jointed, red Wingate Sandstone, slope-forming, purplish Chinle Formation, layered, lighter-red Moenkopi Formation, and white, layered Cutler Formation sandstone. Picture from Glen Canyon National Recreation Area, Utah.

The Principle of Original Horizontality states that layers of sediment are originally deposited horizontally under the action of gravity . It is a relative dating technique. The principle is important to the analysis of folded and tilted strata. It was first proposed by the Danish geological pioneer Nicholas Steno (1638–1686).

From these observations is derived the conclusion that the Earth has not been static and that great forces have been at work over long periods of time, further leading to the conclusions of the science of plate tectonics; that movement and collisions of large plates of the Earth's crust is the cause of folded strata.

As one of Steno's Laws, the Principle of Original Horizontality served well in the nascent days of geological science. However, it is now known that not all sedimentary layers are deposited purely horizontally. For instance, coarser grained sediments such as sand may be deposited at angles of up to 15 degrees, held up by the internal friction between grains which prevents them slumping to a lower angle without additional reworking or effort. This is known as the angle of repose, and a prime example is the surface of sand dunes.

Similarly, sediments may drape over a pre-existing inclined surface: these sediments are usually deposited conformably to the pre-existing surface. Also sedimentary beds may pinch out along strike, implying that slight angles existed during their deposition. Thus the Principle of Original Horizontality is widely, but not universally, applicable in the study of sedimentology, stratigraphy and structural geology.

Cross-cutting Relationships

Cross-cutting relationships is a principle of geology that states that the geologic feature which cuts another is the younger of the two features. It is a relative dating technique in geology. It was first developed by Danish geological pioneer Nicholas Steno in *Dissertationis prodromus* (1669) and later formulated by James Hutton in *Theory of the Earth* (1795) and embellished upon by Charles Lyell in *Principles of Geology* (1830).

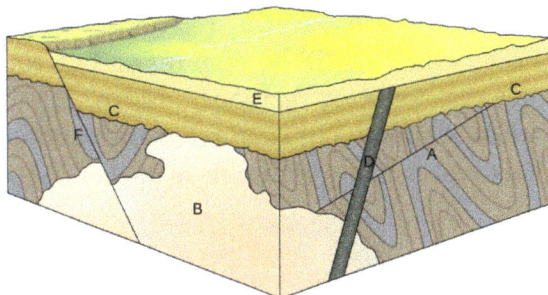

Cross-cutting relations can be used to determine the relative ages of rock strata and other geological structures. Explanations: A - folded rock strata cut by a thrust fault; B - large intrusion (cutting through A); C - erosional angular unconformity (cutting off A & B) on which rock strata were deposited; D - volcanic dike (cutting through A, B & C); E - even younger rock strata (overlying C & D); F - normal fault (cutting through A, B, C & possibly E).

Types

There are several basic types of cross cutting relationships:

- Structural relationships may be faults or fractures cutting through an older rock.

- Intrusional relationships occur when an igneous pluton or dike is intruded into pre-existing rocks.

- Stratigraphic relationships may be an erosional surface (or unconformity) cuts across older rock layers, geological structures, or other geological features.

- Sedimentological relationships occur where currents have eroded or scoured older sediment in a local area to produce, for example, a channel filled with sand.

- Paleontological relationships occur where animal activity or plant growth produces truncation. This happens, for example, where animal burrows penetrate into pre-existing sedimentary deposits.

- Geomorphological relationships may occur where a surficial feature, such as a river, flows through a gap in a ridge of rock. In a similar example, an impact crater excavates into a subsurface layer of rock.

Cross-cutting relationships may be compound in nature. For example, if a fault were truncated by an unconformity, and that unconformity cut by a dike. Based upon such compound cross-cutting relationships it can be seen that the fault is older than the unconformity which in turn is older than the dike. Using such rationale, the sequence of geological events can be better understood.

Scale

Cross-cutting relationships may be seen cartographically, megascopically, and microscopically. In other words, these relationships have various scales. A cartographic crosscutting relationship might look like, for example, a large fault dissecting the landscape on a large map. Megascopic cross-cutting relationships are features like igneous dikes, as mentioned above, which would be seen on an outcrop or in a limited geographic area. Microscopic cross-cutting relationships are those that require study by magnification or other close scrutiny. For example, penetration of a fossil shell by the drilling action of a boring organism is an example of such a relationship.

Cross-cutting relationships involving an andesitic dike in Peru that cuts across the lower sedimentary strata. Both the dike and the lower strata are cut by an unconformity

A light-gray igneous intrusion in Sweden cut by a younger white pegmatite dike, which in turn is cut by an even younger black diabase dike

Other Use

Cross-cutting relationships can also be used in conjunction with radiometric age dating to effect an age bracket for geological materials that cannot be directly dated by radiometric techniques. For example, if a layer of sediment containing a fossil of interest is bounded on the top and bottom by unconformities, where the lower unconformity truncates dike A and the upper unconformity truncates dike B (which penetrates the layer in question), this method can be used. A radiometric age date from crystals in dike A will give the maximum age date for the layer in question and likewise, crystals from dike B will give us the minimum age date. This provides an age bracket, or range of possible ages, for the layer in question.

Historical Development

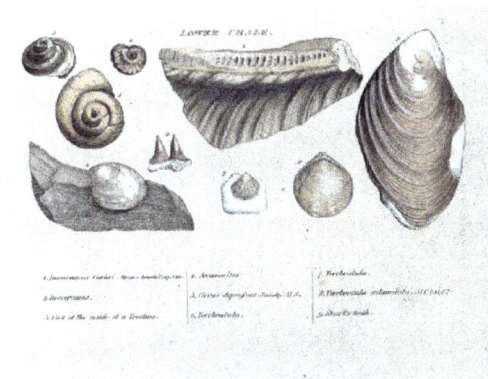

Engraving from William Smith's monograph on identifying strata based on fossils

Nicholas Steno established the theoretical basis for stratigraphy when he introduced the law of superposition, the principle of original horizontality and the principle of lateral continuity in a 1669 work on the fossilization of organic remains in layers of sediment.

The first practical large-scale application of stratigraphy was by William Smith in the 1790s and early 19th century. Smith, known as the "Father of English geology", created the first geologic map of England and first recognized the significance of strata or rock layering and the importance of fossil markers for correlating strata. Another influential application of stratigraphy in the early 19th century was a study by Georges Cuvier and Alexandre Brongniart of the geology of the region around Paris.

Strata in Cafayate (Argentina)

Lithostratigraphy

Chalk layers in Cyprus, showing sedimentary layering

Variation in rock units, most obviously displayed as visible layering, is due to physical contrasts in rock type (lithology). This variation can occur vertically as layering (bedding), or laterally, and reflects changes in environments of deposition (known as facies change). These variations provide a lithostratigraphy or lithologic stratigraphy of the rock unit. Key concepts in stratigraphy involve understanding how certain geometric relationships between rock layers arise and what these geometries imply about their original depositional environment. The basic concept in stratigraphy, called the law of superposition, states: in an undeformed stratigraphic sequence, the oldest strata occur at the base of the sequence.

Chemostratigraphy studies the changes in the relative proportions of trace elements and isotopes within and between lithologic units. Carbon and oxygen isotope ratios vary with time, and researchers can use those to map subtle changes that occurred in the paleoenvironment. This has led to the specialized field of isotopic stratigraphy.

Cyclostratigraphy documents the often cyclic changes in the relative proportions of minerals (particularly carbonates), grain size, thickness of sediment layers (varves) and fossil diversity with time, related to seasonal or longer term changes in palaeoclimates.

Biostratigraphy

Biostratigraphy or paleontologic stratigraphy is based on fossil evidence in the rock layers. Strata from widespread locations containing the same fossil fauna and flora are said to be correlatable in time. Biologic stratigraphy was based on William Smith's principle of faunal succession, which predated, and was one of the first and most powerful lines of evidence for, biological evolution. It provides strong evidence for the formation (speciation) and extinction of species. The geologic time scale was developed during the 19th century, based on the evidence of biologic stratigraphy and faunal succession. This timescale remained a relative scale until the development of radiometric dating, which gave it and the stratigraphy it was based on an absolute time framework, leading to the development of chronostratigraphy.

One important development is the Vail curve, which attempts to define a global historical sea-level curve according to inferences from worldwide stratigraphic patterns. Stratigraphy is also commonly used to delineate the nature and extent of hydrocarbon-bearing reservoir rocks, seals, and traps of petroleum geology.

Chronostratigraphy

Chronostratigraphy is the branch of stratigraphy that places an absolute age, rather than a relative age on rock strata. The branch is concerned with deriving geochronological data for rock units, both directly and inferentially, so that a sequence of time-relative events that created the rocks formation can be derived. The ultimate aim of chronostratigraphy is to place dates on the sequence of deposition of all rocks within a geological region, and then to every region, and by extension to provide an entire geologic record of the Earth.

A gap or missing strata in the geological record of an area is called a stratigraphic hiatus. This may be the result of a halt in the deposition of sediment. Alternatively, the gap may be due to removal by erosion, in which case it may be called a stratigraphic vacuity. It is called a *hiatus* because deposition was *on hold* for a period of time. A physical gap may represent both a period of non-deposition and a period of erosion. A geologic fault may cause the appearance of a hiatus.

Magnetostratigraphy

Magnetostratigraphy is a chronostratigraphic technique used to date sedimentary and volcanic sequences. The method works by collecting oriented samples at measured intervals throughout a section. The samples are analyzed to determine their detrital remanent magnetism (DRM), that is, the polarity of Earth's magnetic field at the time a stratum was deposited. For sedimentary rocks, this is possible because, when very fine-grained magnetic minerals (< 17 µm), behave like tiny compasses as they fall through the water column and orient themselves with Earth's magnetic field. Upon burial, that orientation is preserved. For volcanic rocks, magnetic minerals, which form in the melt, orient themselves with the ambient magnetic field, and are fixed in place upon crystallization of the lava.

Oriented paleomagnetic core samples are collected in the field; mudstones, siltstones, and very fine-grained sandstones are the preferred lithologies because the magnetic grains are finer and more likely to orient with the ambient field during deposition. If the ancient magnetic field were oriented similar to today's field (North Magnetic Pole near the North Rotational Pole), the strata would retain a normal polarity. If the data indicate that the North Magnetic Pole were near the South Rotational Pole, the strata would exhibit reversed polarity.

Results of the individual samples are analyzed by removing the natural remanent magnetization (NRM) to reveal the DRM. Following statistical analysis, the results are used to generate a local magnetostratigraphic column that can then be compared against the Global Magnetic Polarity Time Scale.

This technique is used to date sequences that generally lack fossils or interbedded igneous rocks. The continuous nature of the sampling means that it is also a powerful technique for the estimation of sediment-accumulation rates.

Archaeological Stratigraphy

In the field of archaeology, soil stratigraphy is used to better understand the processes that form and protect archaeological sites. Since the law of superposition holds true, it can help date finds or features from each context; these finds and features can be placed in sequence and the dates interpolated. Phases of activity can also often be seen through stratigraphy, especially when a trench or feature is viewed in section (profile). Because pits and other features can be dug down into earlier levels, not all material at the same absolute depth is necessarily of the same age; close attention has to be paid to the archeological layers. The Harris-matrix is a tool to depict complex stratigraphic relations when they are found, for example, in the context of urban archaeology.

References

- Davies G.L.H. (2007). Whatever is Under the Earth the Geological Society of London 1807-2007. London: Geological Society. p. 78. ISBN 9781862392144.

- Kearey, Philip (2001). Dictionary of Geology (2nd ed.) London, New York, etc.: Penguin Reference, London, p. 123. ISBN 978-0-14-051494-0.

- Chapman, Richard E. (1983) Petroleum Geology Elsevier Scientific, Amsterdam, page 33, ISBN 978-0-444-42165-4

- Principles of Archaeological Stratigraphy. 40 figs. 1 pl. 136 pp. London & New York: Academic Press ISBN 0-12-326650-5

- Levin, H.L. (2009). The Earth Through Time (9 ed.). John Wiley and Sons. p. 15. ISBN 978-0-470-38774-0. Retrieved 28 November 2010.

7

Applications of Sedimentary Rocks

The applications of sedimentary rocks are ceramics, dimension stones, petroleum geology and aquifer. Ceramic is an inorganic material that encompasses of metal or nonmetal atoms. These atoms are held in ionic and covalent bonds. This section has been carefully written to provide an easy understanding of the applications of sedimentary rocks.

Ceramic

A ceramic is an inorganic, nonmetallic solid material comprising metal, nonmetal or metalloid atoms primarily held in ionic and covalent bonds.

A Ming Dynasty porcelain vase dated to 1403–1424

The crystallinity of ceramic materials ranges from highly oriented to semi-crystalline, and often completely amorphous (e.g., glasses). Varying crystallinity and electron consumption in the ionic and covalent bonds cause most ceramic materials to be good thermal and electrical insulators (extensively researched in ceramic engineering). With such a large range of possible options for the composition/structure of a ceramic (e.g. nearly all of the elements, nearly all types of bonding, and all levels of crystallinity), the breadth of the subject is vast, and identifiable attributes (e.g. hardness, toughness, electrical conductivity, etc.) are hard to specify for the group as a whole. General properties such as high melting temperature, high hardness, poor conductivity, high moduli of elasticity, chemical resistance and low ductility are the norm, with known exceptions to each of these rules (e.g. piezoelectric ceramics, glass transition temperature, superconductive ceramics, etc.). Many composites, such as fiberglass and carbon fiber, while containing ceramic materials, are not considered to be part of the ceramic family.

A selection of silicon nitride components.

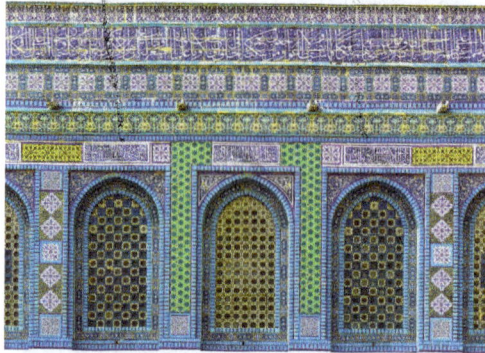

Mid-16th century ceramic tilework on the Dome of the Rock, Jerusalem

The earliest ceramics made by humans were pottery objects (i.e. *pots* or *vessels*) or figurines made from clay, either by itself or mixed with other materials like silica, hardened, sintered, in fire. Later ceramics were glazed and fired to create smooth, colored surfaces, decreasing porosity through the use of glassy, amorphous ceramic coatings on top of the crystalline ceramic substrates. Ceramics now include domestic, industrial and building products, as well as a wide range of ceramic art. In the 20th century, new ceramic materials were developed for use in advanced ceramic engineering, such as in semiconductors.

Spherical Hanging Ornament, 1575-1585, Ottoman Period. Brooklyn Museum.

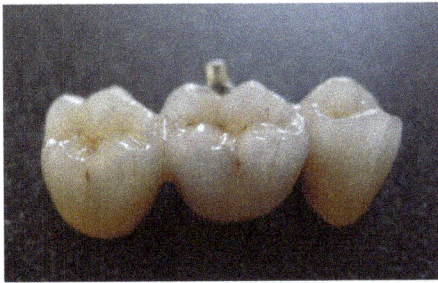

Fixed partial porcelain denture, or "bridge"

The earliest known mention of the root "ceram-" is the "workers of ceramics", written in Linear B syllabic script. The word "ceramic" may be used as an adjective to describe a material, product or process, or it may be used as a noun, either singular, or, more commonly, as the plural noun "ceramics".

Types of Ceramic Material

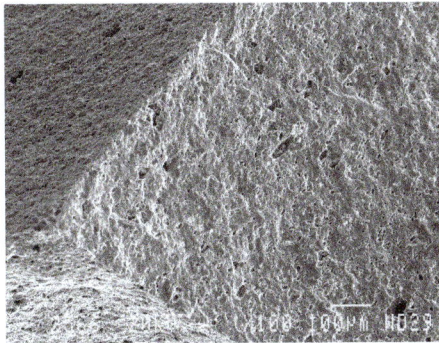

A low magnification SEM micrograph of an advanced ceramic material. The properties of ceramics make fracturing an important inspection method.

A ceramic material is an inorganic, non-metallic, often crystalline oxide, nitride or carbide material. Some elements, such as carbon or silicon, may be considered ceramics.

Ceramic materials are brittle, hard, strong in compression, weak in shearing and tension. They withstand chemical erosion that occurs in other materials subjected to acidic or caustic environments. Ceramics generally can withstand very high temperatures, such as temperatures that range from 1,000 °C to 1,600 °C (1,800 °F to 3,000 °F). Glass is often not considered a ceramic because of its amorphous (noncrystalline) character. However, glassmaking involves several steps of the ceramic process and its mechanical properties are similar to ceramic materials.

Traditional ceramic raw materials include clay minerals such as kaolinite, whereas more recent materials include aluminium oxide, more commonly known as alumina. The modern ceramic materials, which are classified as advanced ceramics, include silicon carbide and tungsten carbide. Both are valued for their abrasion resistance, and hence find use in applications such as the wear plates of crushing equipment in mining operations. Advanced ceramics are also used in the medicine, electrical, electronics industries and body armor.

Crystalline Ceramics

Crystalline ceramic materials are not amenable to a great range of processing. Methods for dealing with them tend to fall into one of two categories – either make the ceramic in the desired shape, by reaction *in situ*, or by "forming" powders into the desired shape, and then sintering to form a solid body. Ceramic forming techniques include shaping by hand (sometimes including a rotation process called "throwing"), slip casting, tape casting (used for making very thin ceramic capacitors, e.g.), injection molding, dry pressing, and other variations. Details of these processes are described in the two books listed below. A few methods use a hybrid between the two approaches.

Noncrystalline Ceramics

Noncrystalline ceramics, being glass, tend to be formed from melts. The glass is shaped when either fully molten, by casting, or when in a state of toffee-like viscosity, by methods such as blowing into a mold. If later heat treatments cause this glass to become partly crystalline, the resulting material is known as a glass-ceramic, widely used as cook-top and also as a glass composite material for nuclear waste disposal.

Properties of Ceramics

The physical properties of any ceramic substance are a direct result of its crystalline structure and chemical composition. Solid state chemistry reveals the fundamental connection between microstructure and properties such as localized density variations, grain size distribution, type of porosity and second-phase content, which can all be correlated with ceramic properties such as mechanical strength σ by the Hall-Petch equation, hardness, toughness, dielectric constant, and the optical properties exhibited by transparent materials.

Physical properties of chemical compounds which provide evidence of chemical composition include odor, colour, volume, density (mass / volume), melting point, boiling point, heat capacity, physical form at room temperature (solid, liquid or gas), hardness, porosity, and index of refraction.

Ceramography is the art and science of preparation, examination and evaluation of ceramic microstructures. Evaluation and characterization of ceramic microstructures is often implemented on similar spatial scales to that used commonly in the emerging field of nanotechnology: from tens of angstroms (A) to tens of micrometers (μm). This is typically somewhere between the minimum wavelength of visible light and the resolution limit of the naked eye.

The microstructure includes most grains, secondary phases, grain boundaries, pores, micro-cracks, structural defects and hardness microindentions. Most bulk mechanical, optical, thermal, electrical and magnetic properties are significantly affected by the observed microstructure. The fabrication method and process conditions are generally indicated by the microstructure. The root cause of many ceramic failures is evident in the cleaved and polished microstructure. Physical properties which constitute the field of materials science and engineering include the following:

Mechanical Properties

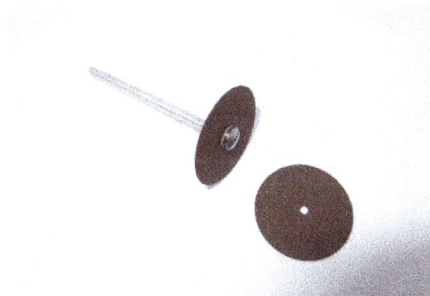

Cutting disks made of silicon carbide

The Porsche Carrera GT's carbon-ceramic (silicon carbide) disc brake

Mechanical properties are important in structural and building materials as well as textile fabrics. They include the many properties used to describe the strength of materials such as: elasticity / plasticity, tensile strength, compressive strength, shear strength, fracture toughness & ductility (low in brittle materials), and indentation hardness.

In modern materials science, fracture mechanics is an important tool in improving the mechanical performance of materials and components. It applies the physics of stress and strain, in particular the theories of elasticity and plasticity, to the microscopic crystallographic defects found in real materials in order to predict the macroscopic mechanical failure of bodies. Fractography is widely used with fracture mechanics to understand the causes of failures and also verify the theoretical failure predictions with real life failures.

Ceramic materials are usually ionic or covalent bonded materials, and can be crystalline or amorphous. A material held together by either type of bond will tend to fracture before any plastic deformation takes place, which results in poor toughness in these materials. Additionally, because these materials tend to be porous, the pores and other microscopic imperfections act as stress concentrators, decreasing the toughness further, and reducing the tensile strength. These combine to give catastrophic failures, as opposed to the normally much more gentle failure modes of metals.

These materials do show plastic deformation. However, due to the rigid structure of the crystalline materials, there are very few available slip systems for dislocations to move, and so they deform very slowly. With the non-crystalline (glassy) materials, viscous flow is the dominant source of plastic deformation, and is also very slow. It is therefore neglected in many applications of ceramic materials.

To overcome the brittle behaviour, ceramic material development has introduced the class of ceramic matrix composite materials, in which ceramic fibers are embedded and with specific coatings are forming fiber bridges across any crack. This mechanism substantially increases the fracture toughness of such ceramics. The ceramic disc brakes are, for example using a ceramic matrix composite material manufactured with a specific process.

Electrical Properties

Semiconductors

Some ceramics are semiconductors. Most of these are transition metal oxides that are II-VI semiconductors, such as zinc oxide.

While there are prospects of mass-producing blue LEDs from zinc oxide, ceramicists are most interested in the electrical properties that show grain boundary effects.

One of the most widely used of these is the varistor. These are devices that exhibit the property that resistance drops sharply at a certain threshold voltage. Once the voltage across the device reaches the threshold, there is a breakdown of the electrical structure in the vicinity of the grain boundaries, which results in its electrical resistance dropping from several megohms down to a few hundred ohms. The major advantage of these is that they can dissipate a lot of energy, and they self-reset – after the voltage across the device drops below the threshold, its resistance returns to being high.

This makes them ideal for surge-protection applications; as there is control over the threshold voltage and energy tolerance, they find use in all sorts of applications. The best demonstration of their ability can be found in electrical substations, where they are employed to protect the infrastructure from lightning strikes. They have rapid response, are low maintenance, and do not appreciably degrade from use, making them virtually ideal devices for this application.

Semiconducting ceramics are also employed as gas sensors. When various gases are passed over a polycrystalline ceramic, its electrical resistance changes. With tuning to the possible gas mixtures, very inexpensive devices can be produced.

Superconductivity

The Meissner effect demonstrated by levitating a magnet above a cuprate superconductor, which is cooled by liquid nitrogen

Under some conditions, such as extremely low temperature, some ceramics exhibit high temperature superconductivity. The exact reason for this is not known, but there are two major families of superconducting ceramics.

Ferroelectricity and Supersets

Piezoelectricity, a link between electrical and mechanical response, is exhibited by a large number of ceramic materials, including the quartz used to measure time in watches and other electronics. Such devices use both properties of piezoelectrics, using electricity to produce a mechanical motion (powering the device) and then using this mechanical motion to produce electricity (generating a signal). The unit of time measured is the natural interval required for electricity to be converted into mechanical energy and back again.

The piezoelectric effect is generally stronger in materials that also exhibit pyroelectricity, and all pyroelectric materials are also piezoelectric. These materials can be used to inter convert between thermal, mechanical, or electrical energy; for instance, after synthesis in a furnace, a pyroelectric crystal allowed to cool under no applied stress generally builds up a static charge of thousands of volts. Such materials are used in motion sensors, where the tiny rise in temperature from a warm body entering the room is enough to produce a measurable voltage in the crystal.

In turn, pyroelectricity is seen most strongly in materials which also display the ferroelectric effect, in which a stable electric dipole can be oriented or reversed by applying an electrostatic field. Pyroelectricity is also a necessary consequence of ferroelectricity. This can be used to store information in ferroelectric capacitors, elements of ferroelectric RAM.

The most common such materials are lead zirconate titanate and barium titanate. Aside from the uses mentioned above, their strong piezoelectric response is exploited in the design of high-frequency loudspeakers, transducers for sonar, and actuators for atomic force and scanning tunneling microscopes.

Positive Thermal Coefficient

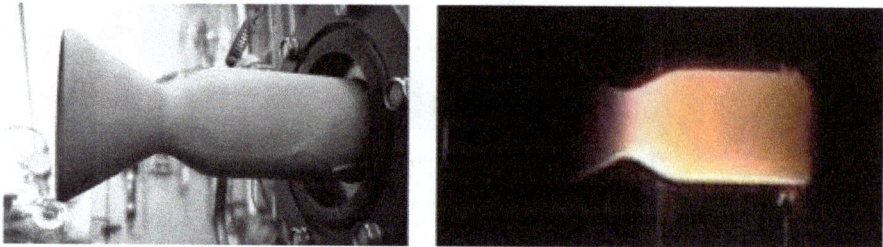

Silicon nitride rocket thruster. Left: Mounted in test stand. Right: Being tested with H_2/O_2 propellants

Increases in temperature can cause grain boundaries to suddenly become insulating in some semiconducting ceramic materials, mostly mixtures of heavy metal titanates. The critical transition temperature can be adjusted over a wide range by variations in chemistry. In such materials, current will pass through the material until joule heating brings it to the transition temperature, at which point the circuit will be broken and current flow will cease. Such ceramics are used as self-controlled heating elements in, for example, the rear-window defrost circuits of automobiles.

At the transition temperature, the material's dielectric response becomes theoretically infinite. While a lack of temperature control would rule out any practical use of the material near its critical temperature, the dielectric effect remains exceptionally strong even at much higher temperatures. Titanates with critical temperatures far below room temperature have become synonymous with "ceramic" in the context of ceramic capacitors for just this reason.

Optical Properties

Cermax xenon arc lamp with synthetic sapphire output window

Optically transparent materials focus on the response of a material to incoming lightwaves of a range of wavelengths. Frequency selective optical filters can be utilized to alter or enhance the brightness and contrast of a digital image. Guided lightwave transmission via frequency selective waveguides involves the emerging field of fiber optics and the ability of certain glassy compositions as a transmission medium for a range of frequencies simultaneously (multi-mode optical fiber) with little or no interference between competing wavelengths or frequencies. This resonant mode of energy and data transmission via electromagnetic (light) wave propagation, though low powered, is virtually lossless. Optical waveguides are used as components in Integrated optical circuits (e.g. light-emitting diodes, LEDs) or as the transmission medium in local and long haul optical communication systems. Also of value to the emerging materials scientist is the sensitivity of materials to radiation in the thermal infrared (IR) portion of the electromagnetic spectrum. This heat-seeking ability is responsible for such diverse optical phenomena as Night-vision and IR luminescence.

Thus, there is an increasing need in the military sector for high-strength, robust materials which have the capability to transmit light (electromagnetic waves) in the visible (0.4 – 0.7 micrometers) and mid-infrared (1 – 5 micrometers) regions of the spectrum. These materials are needed for applications requiring transparent armor, including next-generation high-speed missiles and pods, as well as protection against improvised explosive devices (IED).

In the 1960s, scientists at General Electric (GE) discovered that under the right manufacturing conditions, some ceramics, especially aluminium oxide (alumina), could be made translucent. These translucent materials were transparent enough to be used for containing the electrical plasma generated in high-pressure sodium street lamps. During the past two decades, additional types of transparent ceramics have been developed for applications such as nose cones for heat-seeking missiles, windows for fighter aircraft, and scintillation counters for computed tomography scanners.

In the early 1970s, Thomas Soules pioneered computer modeling of light transmission through translucent ceramic alumina. His model showed that microscopic pores in ceramic, mainly trapped at the junctions of microcrystalline grains, caused light to scatter and prevented true transparency. The volume fraction of these microscopic pores had to be less than 1% for high-quality optical transmission.

This is basically a particle size effect. Opacity results from the incoherent scattering of light at surfaces and interfaces. In addition to pores, most of the interfaces in a typical metal or ceramic object are in the form of grain boundaries which separate tiny regions of crystalline order. When the size of the scattering center (or grain boundary) is reduced below the size of the wavelength of the light being scattered, the scattering no longer occurs to any significant extent.

In the formation of polycrystalline materials (metals and ceramics) the size of the crystalline grains is determined largely by the size of the crystalline particles present in the raw material during formation (or pressing) of the object. Moreover, the size of the grain boundaries scales directly with particle size. Thus a reduction of the original particle size below the wavelength of visible light (~ 0.5 micrometers for shortwave violet) eliminates any light scattering, resulting in a transparent material.

Recently, Japanese scientists have developed techniques to produce ceramic parts that rival the transparency of traditional crystals (grown from a single seed) and exceed the fracture toughness of a single crystal. In particular, scientists at the Japanese firm Konoshima Ltd., a producer of ceramic construction materials and industrial chemicals, have been looking for markets for their transparent ceramics.

Livermore researchers realized that these ceramics might greatly benefit high-powered lasers used in the National Ignition Facility (NIF) Programs Directorate. In particular, a Livermore research team began to acquire advanced transparent ceramics from Konoshima to determine if they could meet the optical requirements needed for Livermore's Solid-State Heat Capacity Laser (SSHCL). Livermore researchers have also been testing applications of these materials for applications such as advanced drivers for laser-driven fusion power plants.

Examples

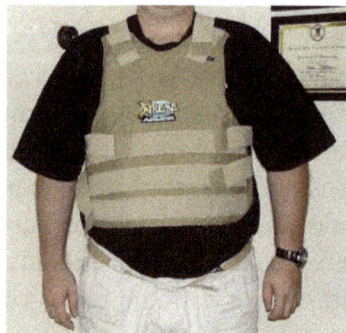

Silicon carbide is used for inner plates of ballistic vests

Porcelain high-voltage insulator

Ceramic BN crucible

Until the 1950s, the most important ceramic materials were (1) pottery, bricks and tiles, (2) cements and (3) glass. A composite material of ceramic and metal is known as cermet.

Other ceramic materials, generally requiring greater purity in their make-up than those above, include forms of several chemical compounds, including:

- Barium titanate (often mixed with strontium titanate) displays ferroelectricity, meaning that its mechanical, electrical, and thermal responses are coupled to one another and also history-dependent. It is widely used in electromechanical transducers, ceramic capacitors, and data storage elements. Grain boundary conditions can create PTC effects in heating elements.

- Bismuth strontium calcium copper oxide, a high-temperature superconductor

- Boron oxide is used in body armor.

- Boron nitride is structurally isoelectronic to carbon and takes on similar physical forms: a graphite-like one used as a lubricant, and a diamond-like one used as an abrasive.

- Earthenware used for domestic ware such as plates and mugs.

- Ferrite is used in the magnetic cores of electrical transformers and magnetic core memory.

- Lead zirconate titanate (PZT) was developed at the United States National Bureau of Standards in 1954. PZT is used as an ultrasonic transducer, as its piezoelectric properties greatly exceed those of Rochelle salt.

- Magnesium diboride (MgB_2) is an unconventional superconductor.

- Porcelain is used for a wide range of household and industrial products.

- Sialon (Silicon Aluminium Oxynitride) has high strength; resistance to thermal shock, chemical and wear resistance, and low density. These ceramics are used in non-ferrous molten metal handling, weld pins and the chemical industry.

- Silicon carbide (SiC) is used as a susceptor in microwave furnaces, a commonly used abrasive, and as a refractory material.

- Silicon nitride (Si_3N_4) is used as an abrasive powder.

- Steatite (magnesium silicates) is used as an electrical insulator.

- Titanium carbide Used in space shuttle re-entry shields and scratchproof watches.

- Uranium oxide (UO_2), used as fuel in nuclear reactors.

- Yttrium barium copper oxide ($YBa_2Cu_3O_{7-x}$), another high temperature superconductor.

- Zinc oxide (ZnO), which is a semiconductor, and used in the construction of varistors.

- Zirconium dioxide (zirconia), which in pure form undergoes many phase changes between room temperature and practical sintering temperatures, can be chemically "stabilized" in several different forms. Its high oxygen ion conductivity recommends it for use in fuel cells and automotive oxygen sensors. In another variant, metastable structures can impart transformation toughening for mechanical applications; most ceramic knife blades are made of this material.

- Partially stabilised zirconia (PSZ) is much less brittle than other ceramics and is used for metal forming tools, valves and liners, abrasive slurries, kitchen knives and bearings subject to severe abrasion.

Kitchen knife with a ceramic blade

Ceramic Products

By Usage

For convenience, ceramic products are usually divided into four main types; these are shown below with some examples:

- Structural, including bricks, pipes, floor and roof tiles

- Refractories, such as kiln linings, gas fire radiants, steel and glass making crucibles

- Whitewares, including tableware, cookware, wall tiles, pottery products and sanitary ware

- Technical, also known as engineering, advanced, special, and fine ceramics. Such items include:

  o gas burner nozzles

  o ballistic protection

  o nuclear fuel uranium oxide pellets

  o biomedical implants

  o coatings of jet engine turbine blades

  o ceramic disk brake

  o missile nose cones

  o bearing (mechanical)

  o tiles used in the Space Shuttle program

Ceramics Made with Clay

Frequently, the raw materials of modern ceramics do not include clays. Those that do are classified as follows:

- Earthenware, fired at lower temperatures than other types

- Stoneware, vitreous or semi-vitreous

- Porcelain, which contains a high content of kaolin

- Bone china

Classification of Technical Ceramics

Technical ceramics can also be classified into three distinct material categories:

- Oxides: alumina, beryllia, ceria, zirconia

- Nonoxides: carbide, boride, nitride, silicide

- Composite materials: particulate reinforced, fiber reinforced, combinations of oxides and nonoxides.

Each one of these classes can develop unique material properties because ceramics tend to be crystalline.

Applications

- Knife blades: the blade of a ceramic knife will stay sharp for much longer than that of a steel knife, although it is more brittle and can snap from a fall onto a hard surface.

- Carbon-ceramic brake disks for vehicles are resistant to brake fade at high temperatures.

- Advanced composite ceramic and metal matrices have been designed for most modern armoured fighting vehicles because they offer superior penetrating resistance against shaped charges (such as HEAT rounds) and kinetic energy penetrators.

- Ceramics such as alumina and boron carbide have been used in ballistic armored vests to repel large-caliber rifle fire. Such plates are known commonly as small arms protective inserts, or SAPIs. Similar material is used to protect the cockpits of some military airplanes, because of the low weight of the material.

- Ceramics can be used in place of steel for ball bearings. Their higher hardness means they are much less susceptible to wear and typically last for triple the lifetime of a steel part. They also deform less under load, meaning they have less contact with the bearing retainer walls and can roll faster. In very high speed applications, heat from friction during rolling can cause problems for metal bearings, which are reduced by the use of ceramics. Ceramics are also more chemically resistant and can be used in wet environments where steel bearings would rust. In some cases, their electricity-insulating properties may also be valuable in bearings. Two drawbacks to ceramic bearings are a significantly higher cost and susceptibility to damage under shock loads.

- In the early 1980s, Toyota researched production of an adiabatic engine using ceramic components in the hot gas area. The ceramics would have allowed temperatures of over 3000 °F (1650 °C). The expected advantages would have been lighter materials and a smaller cooling system (or no need for one at all), leading to a major weight reduction. The expected increase of fuel efficiency of the engine (caused by the higher temperature, as shown by Carnot's theorem) could not be verified experimentally; it was found that the heat transfer on the hot ceramic cylinder walls was higher than the transfer to a cooler metal wall as the cooler gas film on the metal surface works as a thermal insulator. Thus, despite all of these desirable properties, such engines have not succeeded in production because of costs for the ceramic components and the limited advantages. (Small imperfections in the ceramic material with its low fracture toughness lead to cracks, which can lead to potentially dangerous equipment failure.) Such engines are possible in laboratory settings, but mass production is not feasible with current technology.

- Work is being done in developing ceramic parts for gas turbine engines. Currently, even blades made of advanced metal alloys used in the engines' hot section require cooling and careful limiting of operating temperatures. Turbine engines made with ceramics could operate more efficiently, giving aircraft greater range and payload for a set amount of fuel.

- Recent advances have been made in ceramics which include bioceramics, such as dental implants and synthetic bones. Hydroxyapatite, the natural mineral component of bone, has been made synthetically from a number of biological and chemical sources and can be formed into ceramic materials. Orthopedic implants coated with these materials bond readily to bone and other tissues in the body without rejection or inflammatory reactions so are of great interest for gene delivery and tissue engineering scaffolds. Most hydroxyapatite ceramics are very porous and lack mechanical strength, and are used to coat metal orthopedic devices to aid in forming a bond to bone or as bone fillers. They are also used as fillers for orthopedic plastic screws to aid in reducing the inflammation and increase absorption of these plastic materials. Work is being done to make strong, fully dense nanocrystalline hydroxyapatite ceramic materials for orthopedic weight bearing devices, replacing foreign metal and plastic orthopedic materials with a synthetic, but naturally occurring, bone mineral. Ultimately, these ceramic materials may be used as bone replacements or with the incorporation of protein collagens, synthetic bones.

- High-tech ceramic is used in watchmaking for producing watch cases. The material is valued by watchmakers for its light weight, scratch resistance, durability and smooth touch. IWC is one of the brands that initiated the use of ceramic in watchmaking.

Ceramics in Archaeology

Ceramic artifacts have an important role in archaeology for understanding the culture, technology and behavior of peoples of the past. They are among the most common artifacts to be found at an archaeological site, generally in the form of small fragments of broken pottery called sherds. Processing of collected sherds can be consistent with two main types of analysis: technical and traditional.

Traditional analysis involves sorting ceramic artifacts, sherds and larger fragments into specific types based on style, composition, manufacturing and morphology. By creating these typologies it is possible to distinguish between different cultural styles, the purpose of the ceramic and technological state of the people among other conclusions. In addition, by looking at stylistic changes of ceramics over time is it possible to separate (seriate) the ceramics into distinct diagnostic groups (assemblages). A comparison of ceramic artifacts with known dated assemblages allows for a chronological assignment of these pieces.

The technical approach to ceramic analysis involves a finer examination of the composition of ceramic artifacts and sherds to determine the source of the material and through this the possible manufacturing site. Key criteria are the composition of the clay and the temper used in the manufacture of the article under study: temper is a material added to the clay during the initial production stage, and it is used to aid the subsequent drying process. Types of temper include shell pieces, granite fragments and ground sherd pieces called 'grog'. Temper is usually identified by microscopic examination of the temper material. Clay identification is determined by a process of refiring the ceramic, and assigning a color to it using Munsell Soil Color notation. By estimating both the clay and temper compositions, and locating a region where both are known to occur, an assignment of the material source can be made. From the source assignment of the artifact further investigations can be made into the site of manufacture.

Dimension Stone

Large blocks of granite dimension stone being loaded at Teignmouth in 1827

Dimension stone is natural stone or rock that has been selected and finished (i.e., trimmed, cut, drilled, ground, or other) to specific sizes or shapes. Color, texture and pattern, and surface finish of the stone are also normal requirements. Another important selection criterion is durability: the time measure of the ability of dimension stone to endure and to maintain its essential and distinctive characteristics of strength, resistance to decay, and appearance.

Quarries that produce dimension stone or crushed stone (used as construction aggregate) are interconvertible. Since most quarries can produce either one, a crushed stone quarry can be converted to dimension stone production. However, first the stone shattered by heavy and indiscriminate blasting must be removed. Dimension stone is separated by more precise and delicate techniques, such as diamond wire saws, diamond belt saws, burners (jet-piercers), or light and selective blasting with Primacord, a weak explosive.

Stone and Rock Types

A variety of igneous, metamorphic, and sedimentary rocks are used as structural and decorative dimension stone. These rock types are more commonly known as granite, limestone, marble, travertine, quartz-based stone (sandstone, quartzite) and slate. Other varieties of dimension stone that are normally considered to be special minor types include alabaster (massive gypsum), soapstone (massive talc), serpentine and various products fashioned from natural stone.

A variety of finishes can be applied to dimension stone to achieve diverse architectural and aesthetic effects. These finishes include, but are not limited to, the following. A polished finish gives the surface a high luster and strong reflection of incident light (almost mirror-like). A honed finish provides a smooth, satin-like ("eggshell"), non-reflective surface. More textured finishes include brush-hammered, sandblasted, and thermal. A brush-hammered finish, similar to a houndstooth pattern, creates a rough, but uniformly patterned surface with impact tools varying in coarseness. A sandblasted surface provides an irregular pitted surface by impacting sand or metal particles at high velocity against a stone surface. A thermal (or flamed) finish produces a textured, nonreflective surface with only a few reflections from cleavage faces, by applying a high-temperature flame. This finish may change the natural color of the stone depending on mineralogical composition, particularly with stones containing higher levels of iron.

The most easily accessible general (non-graphic) references are the latest Minerals Yearbook Chapter (production and foreign trade, with statistics), and the latest (Issue 31) Dimension Stone Advocate News (new "building green" developments and demand statistics). The most comprehensive, graphic references are Natural Stone Database by Abraxas Verlag (www.natural-stone-database.com), "Dimension Stones of the World, Volumes I & II" (Marble Institute of America) and "Natural Stones Worldwide CD".

Major Applications

While common colors used in some of the major applications are listed below, there is an extraordinarily wide range of colours, available in thousands of patterns. These patterns are created by geological phenomena such as mineral grains, inclusions, veins, cavity fillings, blebs, and streaks. In addition, rocks and stones not normally classed as dimension stone are sometimes selected for these applications. These can included tiles made of jade, agate, and jasper.

Rough cut slabs of granite dimension stone.

Stone (usually granite) countertops and bathroom vanities both involve a finished slab of stone, usually polished but sometimes with another finish (such as honed or sandblasted). Industry standard thicknesses in the United States are 3/4" (2 cm) and 1.25" (3 cm). Often 2 cm slabs will be laminated at the edge to create the appearance of a thicker edge profile. The slabs are cut to fit the top of the kitchen or bathroom cabinet, by measuring, templating or digital templating. Countertop slabs are commonly sawn from rough blocks of stone by reciprocating gangsaws using steel shot as abrasive. More modern technology utilizes diamond wire saws which use less water and energy. Multi-wire saws with as many as 60 wires can slab a block in less than two hours. The slabs are finished (i.e., polished, honed), then sealed with resin to fill micro-fissures and surface imperfections typically due to the loss of poorly bonded elements such as biotite. The fabricators shop cuts these slabs down to final size and finishes the edges with equipment such as hand-held routers, grinders, CNC equipment, or polishers. In 2008, concerns were raised regarding radon emissions from granite countertops; the National Safety Council states that the contributions of radon to inside air come from the soil and rock around the residence (69%), the outdoor air and the water supply (28%), and only 2.5% from all building materials-including granite countertops. A concerned homeowner can employ ASTM radon mitigation and removal techniques. The stone for countertops or vanities is usually granite, but often is marble (especially for vanity tops), and is sometimes limestone or slate. The majority of the stone for this application is produced in Brazil, Italy, and China.

Tile is a thin modular stone unit, commonly 12 in. square (30.5 cm) and 3/8 in. (10 mm) deep. Other popular sizes are 15 in. square (38 cm), 18 in. square (46 cm), and 24 in. square (61 cm); these will usually be deeper than the 12 in. square. The majority of tile has a polished finish, but other finishes such as honed are becoming more common. Almost all stone tile is mass-produced by automated tile lines to identical size, finish, and close tolerances. Exceptions include slate flooring tile and special orders: tile with odd sizes or shapes, unusual finishes, or inlay work. In summary, the automated tile line is a complicated complex of cutting and calibrating machines, honing-polishing machines, edging machines that put on flat or rounded edges, and interconnecting conveyors to move the stone from the slab input to the final tile product. The stone for tiles is most commonly marble, but often is granite, and sometimes limestone, slate, or quartz-based stone. Common colors are white and light earth colors. Much of the stone for this application is produced in Italy and China.

Slate tile covers this entire structure in Germany

Stone monuments include tombstones, grave markers or as mausoleums. After being gangsawed into big deep (up to 10 ft (3.0 m) wide and over 6 inches deep) slabs, smaller saws or guillotines (they break the granite and make the rough edges commonly seen on monuments) shape the monuments. The fronts and backs are usually polished. The individual monuments are then carved, shaped, and further defined by hand tools and sandblasting equipment. At this time, the stone for monuments is most commonly granite, sometimes marble (as in military cemeteries), and rarely others. Granite and quartz both demonstrate good durability, especially because rain is naturally acidic. (This is a natural consequence of the carbon dioxide present in the atmosphere, which generates a weak solution of carbonic acid in rainfall; further acidification of rainfall arises from oxides of sulphur and nitrogen due to anthropogenic emissions). Long-term durability of quartz monuments can be observed at the Harpers Ferry cemetery, where an old rose quartz tombstone stands. (Limestone and sandstone were commonly chosen for monuments in the nineteenth century, but they are no widely longer used because of the rapid erosion rates due to dissolution of acid-vulnerable carbonates by acidic rainfall.) The most common monument colors for granite are gray, black, and mahogany; for marble, white is most popular. Today, the majority of the stone used in

North America in this application is imported from countries such as India and China. This has depressed traditional North American monument centers such as Georgia and Quebec.

Dimension stone has been used in the construction of buildings for centuries. Due to costs, today stone veneers are usually used in place of solid stone blocks. This courthouse was built of dimension stone quarried in Berea, Ohio.

There are a number of smaller applications for buildings and traffic-related uses. Building components include stone used as veneer (exterior), ashlar, or other shapes. Veneer is a non load-bearing facing of stone attached to a backing of an ornamental nature, although it also protects and insulates. Ashlar is a square block of stone, often brick-sized, for facing of walls (primarily exterior). The other shapes are rectangular blocks used for stair treads, sills, and coping (coping is sometimes nonrectangular). The shapes subject to foot traffic will usually have an abrasive finish such as honed or sandblasted. The stone is mostly limestone, but often is quartz-based stone (sandstone), or even marble or granite. Roofing slate is a thin-split shingle-sized piece of slate, and when in place forms the most permanent kind of roof; slate is also used as countertops and flooring tile. Traffic-related stone is that which is used for curbing (vehicular) and flagstone (pedestrian). Curbing is thin stone slabs used along streets or highways to maintain the integrity of sidewalks and borders. Flagstone is a shallow naturally irregular-edged slab of stone, sometimes sawed into a rectangular shape, used as paving (almost always pedestrian). For curbing, the stone is almost always granite, and for flagstone the stone is almost always quartz-based stone (sandstone or quartzite).

There are several other applications resembling flagstone in using rough dimension (or crushed) stone, usually as quarried, sometimes made smaller (i.e. by a jackhammer), often simply put in place: dry stone and riprap.

The stone used in these applications usually has to have certain properties, or meet a standard specification. The American Society for Testing and Materials (ASTM) has such specifications for granite, marble, limestone, quartz-based dimension stone (C616), slate (C629), travertine (C1527), and serpentine (C1526).

Production

Marble quarry in Carrara, Italy

The major producers of dimension stone include Brazil, China, India, Italy, and Spain, and each have annual production levels of nine to over twenty-two million tons. Portugal produces 3 million tons of dimension stone each year.

According to the USGS, 2007 U.S. dimension stone production was 1.39 million tons valued at $275 million, compared to 1.33 million tons (revised) valued at $265 million in 2006. Of these, granite production was 453,000 tons valued at $106 million in 2007 and 428,000 tons valued at $105 million in 2006, and limestone was 493,000 tons valued at $93.3 million in 2007 and 559,000 tons valued at $96.1 million in 2006. The United States is at best a mid-level dimension stone producer on the world scene; Portugal produces twice as much dimension stone annually.

World comparison for dimension stone demand: The DSAN World Demand Index for (finished) Granite was 227 in 2006, 247 in 2007, and 249 in 2008, and the World Demand Index for (finished) Marble was 200 in 2006, 248 in 2007, and 272 in 2008. The DSAN World Demand for (finished) Granite Index showed a growth of 12% annually for the 2000-2008 period, compared to 14% annually for the 2000-2007 period, and compared to 15% annually for the 2000-2006 period. The DSAN World Demand for (finished) Marble Index showed a growth of 13.5% annually for the 2000-2008 period, compared to 14.0% annually for the 2000-2007 period, and compared to 12.5% annually for the 2000-2006 period. The indexes show world demand for granite has clearly been weakening since 2006, while the world demand for marble only weakened from 2007 to 2008. Other DSAN indexes for 2008 indicate that the 2000-2008 growth was down from the 2000-2007 growth.

The DSAN U.S. Ceramic Tile Demand Index shows a drop of 4.8% annually for the 2000-2007 period, compared to growth of 5.0% annually for the 2000-2006 period. The "traditional" major ceramic tile suppliers, Italy and Spain, have been losing markets to new entrants Brazil and China. The same thing has been happening with dimension stone with increasing supplies from Brazil, China and India.

In 2008, Chinese exports of granite countertops and marble tile increased from 2007, while those of Italy and Spain did not In early 2009, the Chinese Government has a hands-off policy towards its dimension stone industry.

"Building green" with Dimension Stone

Marble cladding on a building

Green building or environmentally friendly construction with natural materials, is an idea that has been around for several decades. Energy price increases and the need for energy conservation when heating or cooling buildings have recently brought it to the fore. This resulted in the formation in 1993 of the U.S. Green Building Council (USGBC), which has developed a building rating system called Leadership in Energy and Environmental Design (LEED). Educational institutions (colleges, universities, grade, and high schools) are often requiring new buildings to be green, and a few jurisdictions (i.e., some cities) have some rules promoting green building. When "building green", dimension stone has a big advantage over steel, concrete, glazed glass and laminated plastics, whose productions are all energy intensive and create significant air and water pollution. As an entirely natural product, dimension stone also has an advantage over synthetic/artificial stone products, as well as composite and space-age materials. One LEED requirement provides that the dimension stone used in a green building be quarried within a 500-mile (800 km) radius of the building being constructed. This gives a clear advantage to domestic dimension stone, plus some quarried near the U.S. borders with Canada and Mexico. A current problem is how to consider stone quarried domestically, sent to China or Italy for finishing, and shipped back to be used in a project.

When demolishing a structure, dimension stone is 100% reusable and can be salvaged for new construction, used as paving or crushed for use as aggregates. There are also "green" methods of stone cleaning either in development or already in use, such as removing the black gypsum crusts that form on marble and limestone by applying sulfate-reducing bacteria to the crust to gasify it, breaking up the crust for easy removal.

The Federal Trade Commission (FTC) in America is re-examining and will most likely update its "Green Guides" used to regulate green advertising claims. The updating will emphasize green building, including the products it involves, such as dimension stone. When the new requirements are finalized, the FTC will go after firms that violate the new requirements, in order to establish legal precedents.

The Natural Stone Council has a library of information on building green with dimension stone, including life-cycle inventory data for each major dimension stone, giving the amount of energy, water, other inputs, and processing emissions, plus some best practice studies. In addition, it has shown ways that dimension stone can contribute LEED points, such as using a light-colored dimension stone to reduce heat-island effects, using dimension stone's thermal mass to impact indoor ambient air temperature thereby increasing energy efficiency, and especially by reusing dimension stone rather sending it to the landfill.

Sustainability of Dimension Stone

Dimension stone is one of the most sustainable of the industrial minerals since it is created by separating it from the natural bedrock underlying all land on every continent. Dimension stone rates very well in terms of the criteria on the ASTM checklist for sustainability of building products: there are no toxic materials used in its processing, there are no direct greenhouse gas emissions during processing, the dust created is controlled, the water used is almost completely recycled (per OSHA/ MSHA regulation), and it is a perpetual resource (virtually inexhaustible in a human time scale). Dimension stone in use can last many generations, even centuries, so the dimension stone manufacturers have not needed a product recycling program. However, there are practical qualifications to and constraints on that sustainability. The dimension stone color and pattern can be changed by weathering when it is very near the surface. The color and pattern can also be changed by proximity to an igneous rock body or by the presence of circulating groundwater charged with carbon dioxide (i.e., limestone, travertine, marble). On the other hand, changes in color and/or pattern can be positive. For example, there are at least 14 separately trade-named varieties of Carrara Marble with many patterns (or no patterns) ranging in shade from white to gray. The presence of faults or closely spaced joints can render the stone unusable. These faults and joints do not have to be at odd angles in the stone mass. Closely spaced, wrongly spaced, or nonparallel bedding planes can make the stone unusable, particularly if the bedding planes are planes of weakness. If part of the stone in one area is unusable, there will be another usable part of the stone elsewhere in the formation. A quarry is not a short-term project unless it encounters one of these constraints. Examples of big, old quarries operating for more than a century include the Barre (VT) granite quarry, the Georgia Marble quarry at Tate, several of the Carrara (Italy) marble quarries, and the Penrhyn (Wales) slate quarry. A quarry will produce dust, noise, and some water pollution, but these can be remedied without too much trouble. The landscape may also have to be restored if quarry waste is temporarily or permanently placed on adjacent land.

Stone Recycling and Reuse

Reconstruction of the Charles Bridge in Prague showing numbered dimension blocks.

Recycling dimension stone can occur when structures are demolished, along with recycling timber and recycling construction aggregate in the form of concrete. The material most likely to be recycled is concrete, and this represents the largest volume of recycled construction material. Not too many structures incorporate dimension stone, and even fewer of them have dimension stone worth saving. Stone recycling is usually done by specialists that monitor local demolition activity, looking for stone-containing houses, buildings, bridge abutments, and other dimension stone structures scheduled for demolition. Particularly treasured are old hand-carved stone pieces with the chisel marks still on them, local stones no longer quarried or that are quarried in a different shade of color or appearance. There is no national or regional trade in reclaimed stone, so a large storage yard is required, since the recovered stone may not be quickly sold and reused. The recycled dimension stone is used in old stone buildings being renovated (to replace deteriorated stone pieces), in fireplace mantels, benches, veneer, or for landscaping (like for retaining walls).

The Parthenon in Athens underwent a major reconstruction prior to the 2004 Olympics

Related to stone recycling and stone reuse is the deconstruction and reconstruction of a stone building. The building is taken apart stone block by stone block and the location and orientation of each block is carefully noted. Any roofing slate and interior stone in place is catalogued and moved in the same fashion. After the blocks, slate, and other stone used have been transported to the new location, they are put back in place where and how they were originally, thus reassembling the building. This is typically very expensive and rare but valuable in terms of historic preservation.

Dimension stone is also reused. Buildings immediately spring to mind, but such things as the ornate stone walls, arches, stairways and balustrades alongside a boulevard can also be renovated and reused. Sometimes the old interior of the building is kept as is, after repair. Sometimes the old building is gutted, leaving only a shell or facade and the space inside reconfigured and modernized. The stone work will usually need attention too.

The old stone work may only need cleaning or sandblasting, but it may need more. Firstly, the building exterior (facade) needs to be inspected for unsafe conditions. Next, the building walls need to be inspected for water leakages. The most likely needs are mortar restoration (repointing), applying consolidants to the old stone, or replacing pieces of stone that are deteriorated (damaged) beyond the point of any repair. The repointing is the removal of existing damaged mortar from the outer portion of the joint between stone units and its replacement by new mortar matching the appearance of the old. The consolidants re-establish the original natural bonding between the stone particles that weathering has removed. Deteriorated pieces of stone work are replaced with pieces of stone that match the original as much as possible. Exterior dimension stone will often change color after exposure to weather over time. For example, Indiana Limestone will weather from a tan to an attractive light yellow. Interior dimension stone can sometimes change its shade a little over time too. For both, it may not be possible to find an exact match, even from the original quarry. Stone will often change its appearance from location to location in the same quarry. If the dimension stone renovationist is truly fortunate, the original builder put aside some spare pieces of the stone for future need.

Stone: Life-cycle Assessment and Best Practices

As in every economic sector, the construction industry's purchases of materials and services creates a whole chain of processes from raw material selecting in situ, removal from the earth, usually proceeding to cutting, finishing, or processing/manufacturing, then transport, and retailing. All of these activities have significant upstream (off-site) environmental impacts, whether in terms of energy and raw resource use or emissions to air, land, or water impacting living organisms or the Earth's surface (non-organic). Life cycle assessment is a method for estimating and comparing a range of environmental performance measures (e.g. global warming, acidification potential, toxicity, ozone depletion potentials) over the full life cycle of a product, a building assembly, or a whole building. As such, it provides a comprehensive means for evaluating and comparing products rather than prescriptive measures of individual product characteristics.

The ASTM has some relevant standards, particularly a guide on environmental life cycle assessment of building materials/products (E1991) that shows how to minimize the subjectivity that commonly mars and confuses environmental decision making. In particular, this guide describes the inventory analysis phase that requires data that is suitable for its intended purpose, thus covering data quality (such as completeness, reliability, accuracy, and credibility) as well as the allocation of the data (for multiple inputs and outputs), among other things. Results have to be on a common basis to allow a statistically significant comparison of alternative building product differences in the interpretation.

The Natural Stone Council (NSC) has commissioned some life-cycle inventory data for use in life cycle assessments. Almost 90% of the effort in doing a life cycle assessment involves getting reliable data. For example, the NSC has data that the Global Warming Potential for granite quarrying is 100 kg of carbon dioxide equivalents and for granite processing is 500 (same units); and the Global Warming Potential for limestone quarrying is 20 kg carbon dioxide equivalents while for limestone processing it is 80 (same units). The data on energy and water use include everything back to removal of overburden in the dimension stone quarry and upstream production of energy and fuels, and forward to packaging of finished dimension stone product or slabs for shipment and transport, or to moving scrap stone to storage or reclamation and to capturing and treatment of dust and waste water. The data is then placed in an impact category (i.e. changes to air, changes to water), characterized as to the contribution of the item to the impact compared to other items, and then the impact categories are assigned weights among themselves to show their relative importance.

The Natural Stone Council has also commissioned four Best Practices. One is on water consumption, treatment, and reuse while extracting and processing dimension stone, including dust mitigation, sludge management, and maximizing water recycling. Another is on site maintenance and quarry closure, including minimizing dust, noise, vibration and keeping the operation clean and tidy, both of which help in restoring the surface upon quarry closure. A third one is on solid waste management, including overburden, damaged stone unsaleable as product, sludge deposited from waste water, spent or spilled petroleum products, or metal scrap. The fourth one is on efficiently transporting stone to be finished as products, then transporting the products to consumers by centralizing freight management, consolidating small loads, choosing appropriate trucks, balancing and securing the load, and packaging with sustainable materials.

Stone Selection and Cleaning

The selector of dimension stone begins by considering stone color and appearance, and how the stone will match its surroundings. The selector has literally thousands of options to choose from, and should examine many options. In addition to many hundreds of different stones with different colors and patterns, each stone can change radically in color and appearance when a different finish is put on it. A polished finish accentuates the color and makes any pattern more vivid, and the rougher finishes (i.e. honed, thermal) lighten the color and make the patterns more subdued. With thousands of possibilities, the selector must start by looking at many stones in many different finishes, or photos of them. Such photos can be found on some dimension stone websites, and on DSAN's Architects Stone Selection Helper.

In addition to selecting a stone color and pattern, the suitability of its properties for the intended use must be considered. Stone being chosen for countertops or vanities should be nonabsorptive, resist stains, and be heat and impact resistant. Stone being used in tiles should be sealed in order to resist staining by spilled liquids. Stone being used for flooring, paving, or surfaces subject to foot or vehicular traffic ought to have a semiabrasive finish for slip resistance, such as bush-hammered or thermal. A glossy polished finish will be slick. Most flagstone surfaces are rough enough to be naturally slip-resistant.

Dimension stone requires some specialized methods for cleaning and maintenance. Abrasive cleaners should not be used on a polished stone finish because it will wear the polish off. Acidic cleaners can not be used on marble or limestone because it will remove (i.e. dissolve) the finish. Textured finishes (thermal, bush-hammered) can be treated with some mildly abrasive cleaners but not bleach or an acidic cleaner (if marble or limestone). Stains are another consideration; stains can be organic (food, grease, or oil) or metallic (iron, copper). Stains require some special removal techniques, such as the poultice method. A new method of cleaning stone on ancient buildings (medieval and renaissance) has been developed in Europe: sulfur-reducing bacteria are used on the black gypsum-containing crusts that form on such buildings to convert the sulfur to a gas that dissipates, thus destroying the crust while leaving the patina produced by aging on the underlying stone. This method is still in development and not yet commercially available.

Stone Finishes

The surface of a stone may be finished in a variety of ways. Below are some typical terms:

- Polished finish - a glossy surface which brings out the full color and character of the stone.

- Hone finish - a satin smooth surface finish with little or no gloss. This is recommended for commercial floors.

- Thermal finish - a surface treatment applied by intense heat flaming.

- Diamond sawed - finish produced by sawing with a diamond toothed saw.

- Rough sawn - a surface finish resulting from the gang sawing (or frame saw) process.

- Brush-hammered - a mechanical process which produces textured surfaces. Texture varies from subtle to rough.

Petroleum Geology

Petroleum geology is the study of origin, occurrence, movement, accumulation, and exploration of hydrocarbon fuels. It refers to the specific set of geological disciplines that are applied to the search for hydrocarbons (oil exploration).

Sedimentary Basin Analysis

Petroleum geology is principally concerned with the evaluation of seven key elements in sedimentary basins:

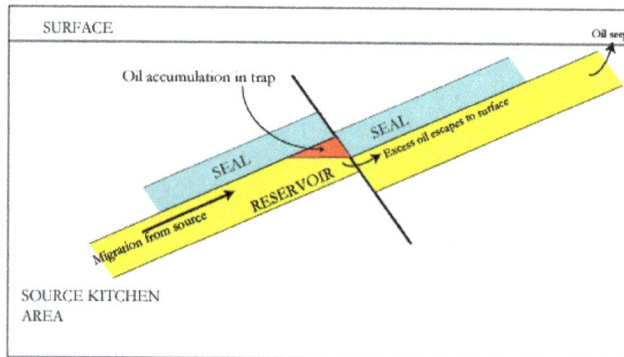

A structural trap, where a fault has juxtaposed a porous and permeable reservoir against an impermeable seal. Oil (shown in red) accumulates against the seal, to the depth of the base of the seal. Any further oil migrating in from the source will escape to the surface and seep.

- Source
- Reservoir
- Seal
- Trap
- Timing
- Maturation
- Migration

In general, all these elements must be assessed via a limited 'window' into the subsurface world, provided by one (or possibly more) exploration wells. These wells present only a 1-dimensional segment through the Earth and the skill of inferring 3-dimensional characteristics from them is one of the most fundamental in petroleum geology. Recently, the availability of inexpensive, high quality 3D seismic data (from reflection seismology) and data from various electromagnetic geophysical techniques (such as Magnetotellurics) has greatly aided the accuracy of such interpretation. The following section discusses these elements in brief. For a more in-depth treatise.

Evaluation of the source uses the methods of geochemistry to quantify the nature of organic-rich rocks which contain the precursors to hydrocarbons, such that the type and quality of expelled hydrocarbon can be assessed.

The reservoir is a porous and permeable lithological unit or set of units that holds the hydrocarbon reserves. Analysis of reservoirs at the simplest level requires an assessment of their porosity (to calculate the volume of *in situ* hydrocarbons) and their permeability (to calculate how easily hydrocarbons will flow out of them). Some of the key disciplines used in reservoir analysis are the fields of structural analysis, stratigraphy, sedimentology, and reservoir engineering.

The seal, or cap rock, is a unit with low permeability that impedes the escape of hydrocarbons from the reservoir rock. Common seals include evaporites, chalks and shales. Analysis of seals involves assessment of their thickness and extent, such that their effectiveness can be quantified.

The trap is the stratigraphic or structural feature that ensures the juxtaposition of reservoir and seal such that hydrocarbons remain trapped in the subsurface, rather than escaping (due to their natural buoyancy) and being lost.

Analysis of maturation involves assessing the thermal history of the source rock in order to make predictions of the amount and timing of hydrocarbon generation and expulsion.

Finally, careful studies of migration reveal information on how hydrocarbons move from source to reservoir and help quantify the source (or *kitchen*) of hydrocarbons in a particular area.

Mud log in process, a common way to study the lithology when drilling oil wells.

Major Subdisciplines in Petroleum Geology

Several major subdisciplines exist in petroleum geology specifically to study the seven key elements discussed above.

Source Rock Analysis

In terms of source rock analysis, several facts need to be established. Firstly, the question of whether there actually *is* any source rock in the area must be answered. Delineation and identification of potential source rocks depends on studies of the local stratigraphy, palaeogeography and sedimentology to determine the likelihood of organic-rich sediments having been deposited in the past.

If the likelihood of there being a source rock is thought to be high, the next matter to address is the state of thermal maturity of the source, and the timing of maturation. Maturation of source rocks depends strongly on temperature, such that the majority of oil generation occurs in the 60° to 120°C range. Gas generation starts at similar temperatures, but may continue up beyond this range, perhaps as high as 200°C. In order to determine the likelihood of oil/gas generation, therefore, the thermal history of the source rock must be calculated. This is performed with a combination of geochemical analysis of the source rock (to determine the type of kerogens present and their maturation characteristics) and basin modelling methods, such as back-stripping, to model the thermal gradient in the sedimentary column.

Basin Analysis

A full scale basin analysis is usually carried out prior to defining leads and prospects for future drilling. This study tackles the petroleum system and studies source rock (presence and quality); burial history; maturation (timing and volumes); migration and focus; and potential regional seals and major reservoir units (that define carrier beds). All these elements are used to investigate where potential hydrocarbons might migrate towards. Traps and potential leads and prospects are then defined in the area that is likely to have received hydrocarbons.

Exploration Stage

Although a basin analysis is usually part of the first study a company conducts prior to moving into an area for future exploration, it is also sometimes conducted during the exploration phase. Exploration geology comprises all the activities and studies necessary for finding new hydrocarbon occurrence. Usually seismic (or 3D seismic) studies are shot, and old exploration data (seismic lines, well logs, reports) are used to expand upon the new studies. Sometimes gravity and magnetic studies are conducted, and oil seeps and spills are mapped to find potential areas for hydrocarbon occurrences.

As soon as a significant hydrocarbon occurrence is found by an exploration- or wild-cat-well the appraisal stage starts.

Appraisal Stage

The Appraisal stage is used to delineate the extent of the discovery. Hydrocarbon reservoir properties, connectivity, hydrocarbon type and gas-oil and oil-water contacts are determined to calculate potential recoverable volumes. This is usually done by drilling more appraisal wells around the initial exploration well. Production tests may also give insight in reservoir pressures and connectivity. Geochemical and petrophysical analysis gives information on the type (viscosity, chemistry, API, carbon content, etc.) of the hydrocarbon and the nature of the reservoir (porosity, permeability, etc.).

Production Stage

After a hydrocarbon occurrence has been discovered and appraisal has indicated it is a commercial find the production stage is initiated. This stage focuses on extracting the hydrocarbons in a controlled way (without damaging the formation, within commercial favorable volumes, etc.). Production wells are drilled and completed in strategic positions. 3D seismic is usually available by this stage to target wells precisely for optimal recovery. Sometimes enhanced recovery (steam injection, pumps, etc.) is used to extract more hydrocarbons or to redevelop abandoned fields.

Reservoir Analysis

The existence of a reservoir rock (typically, sandstones and fractured limestones) is determined through a combination of regional studies (i.e. analysis of other wells in the area), stratigraphy and sedimentology (to quantify the pattern and extent of sedimentation) and seismic interpretation. Once a possible hydrocarbon reservoir is identified, the key physical characteristics of a reservoir that are of interest to a hydrocarbon explorationist are its bulk rock volume, net-to-gross ratio, porosity and permeability.

Bulk rock volume, or the gross rock volume of rock above any hydrocarbon-water contact, is determined by mapping and correlating sedimentary packages. The net-to-gross ratio, typically estimated from analogues and wireline logs, is used to calculate the proportion of the sedimentary packages that contains reservoir rocks. The bulk rock volume multiplied by the net-to-gross ratio gives the net rock volume of the reservoir. The net rock volume multiplied by porosity gives the total hydrocarbon pore volume i.e. the volume within the sedimentary package that fluids (importantly, hydrocarbons and water) can occupy. The summation of these volumes for a given exploration prospect will allow explorers and commercial analysts to determine whether a prospect is financially viable.

Traditionally, porosity and permeability were determined through the study of drilling samples, analysis of cores obtained from the wellbore, examination of contiguous parts of the reservoir that outcrop at the surface and by the technique of formation evaluation using wireline tools passed down the well itself. Modern advances in seismic data acquisition and processing have meant that seismic attributes of subsurface rocks are readily available and can be used to infer physical/sedimentary properties of the rocks themselves.

Aquifer

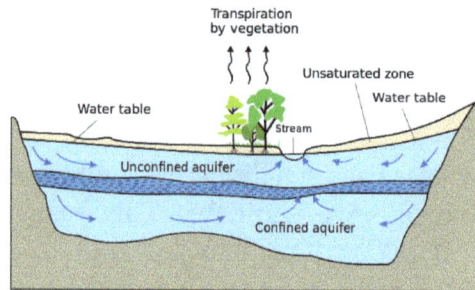

Typical aquifer cross-section

An aquifer is an underground layer of water-bearing permeable rock, rock fractures or unconsolidated materials (gravel, sand, or silt) from which groundwater can be extracted using a water well. The study of water flow in aquifers and the characterization of aquifers is called hydrogeology. Related terms include aquitard, which is a bed of low permeability along an aquifer, and aquiclude (or *aquifuge*), which is a solid, impermeable area underlying or overlying an aquifer. If the impermeable area overlies the aquifer, pressure could cause it to become a confined aquifer.

Depth

Aquifers may occur at various depths. Those closer to the surface are not only more likely to be used for water supply and irrigation, but are also more likely to be topped up by the local rainfall. Many desert areas have limestone hills or mountains within them or close to them that can be exploited as groundwater resources. Part of the Atlas Mountains in North Africa, the Lebanon and Anti-Lebanon ranges between Syria and Lebanon, the Jebel Akhdar (Oman) in Oman, parts of the Sierra Nevada and neighboring ranges in the United States' Southwest, have shallow aquifers that are exploited for their water. Overexploitation can lead to the exceeding of the practical sustained yield;

i.e., more water is taken out than can be replenished. Along the coastlines of certain countries, such as Libya and Israel, increased water usage associated with population growth has caused a lowering of the water table and the subsequent contamination of the groundwater with saltwater from the sea.

The beach provides a model to help visualize an aquifer. If a hole is dug into the sand, very wet or saturated sand will be located at a shallow depth. This hole is a crude well, the wet sand represents an aquifer, and the level to which the water rises in this hole represents the water table.

In 2013 large freshwater aquifers were discovered under continental shelves off Australia, China, North America and South Africa. They contain an estimated half a million cubic kilometers of "low salinity" water that could be economically processed into potable water. The reserves formed when ocean levels were lower and rainwater made its way into the ground in land areas that were not submerged until the ice age ended 20,000 years ago. The volume is estimated to be 100x the amount of water extracted from other aquifers since 1900.

Classification

The above diagram indicates typical flow directions in a cross-sectional view of a simple confined or unconfined aquifer system. The system shows two aquifers with one aquitard (a confining or impermeable layer) between them, surrounded by the bedrock *aquiclude*, which is in contact with a gaining stream (typical in humid regions). The water table and unsaturated zone are also illustrated. An *aquitard* is a zone within the earth that restricts the flow of groundwater from one aquifer to another. An aquitard can sometimes, if completely impermeable, be called an *aquiclude* or *aquifuge*. Aquitards are composed of layers of either clay or non-porous rock with low hydraulic conductivity.

Saturated Versus Unsaturated

Groundwater can be found at nearly every point in the Earth's shallow subsurface to some degree, although aquifers do not necessarily contain fresh water. The Earth's crust can be divided into two regions: the *saturated zone* or *phreatic zone* (e.g., aquifers, aquitards, etc.), where all available spaces are filled with water, and the *unsaturated zone* (also called the vadose zone), where there are still pockets of air that contain some water, but can be filled with more water.

Saturated means the pressure head of the water is greater than atmospheric pressure (it has a gauge pressure > 0). The definition of the water table is the surface where the pressure head is equal to atmospheric pressure (where gauge pressure = 0).

Unsaturated conditions occur above the water table where the pressure head is negative (absolute pressure can never be negative, but gauge pressure can) and the water that incompletely fills the pores of the aquifer material is under suction. The water content in the unsaturated zone is held in place by surface adhesive forces and it rises above the water table (the zero-gauge-pressure isobar) by capillary action to saturate a small zone above the phreatic surface (the capillary fringe) at less than atmospheric pressure. This is termed tension saturation and is not the same as saturation on a water-content basis. Water content in a capillary fringe decreases with increasing distance from the phreatic surface. The capillary head depends on soil pore size. In sandy soils with larger pores, the head will be less than in clay soils with very small pores. The normal capillary rise in a clayey soil is less than 1.80 m (six feet) but can range between 0.3 and 10 m (one and 30 ft).

The capillary rise of water in a small-diameter tube involves the same physical process. The water table is the level to which water will rise in a large-diameter pipe (e.g., a well) that goes down into the aquifer and is open to the atmosphere.

Aquifers Versus Aquitards

Aquifers are typically saturated regions of the subsurface that produce an economically feasible quantity of water to a well or spring (e.g., sand and gravel or fractured bedrock often make good aquifer materials).

An aquitard is a zone within the earth that restricts the flow of groundwater from one aquifer to another. A completely impermeable aquitard is called an aquiclude or aquifuge. Aquitards comprise layers of either clay or non-porous rock with low hydraulic conductivity.

In mountainous areas (or near rivers in mountainous areas), the main aquifers are typically unconsolidated alluvium, composed of mostly horizontal layers of materials deposited by water processes (rivers and streams), which in cross-section (looking at a two-dimensional slice of the aquifer) appear to be layers of alternating coarse and fine materials. Coarse materials, because of the high energy needed to move them, tend to be found nearer the source (mountain fronts or rivers), whereas the fine-grained material will make it farther from the source (to the flatter parts of the basin or overbank areas - sometimes called the pressure area). Since there are less fine-grained deposits near the source, this is a place where aquifers are often unconfined (sometimes called the forebay area), or in hydraulic communication with the land surface.

Confined Versus Unconfined

There are two end members in the spectrum of types of aquifers; *confined* and *unconfined* (with semi-confined being in between). Unconfined aquifers are sometimes also called *water table* or *phreatic* aquifers, because their upper boundary is the water table or phreatic surface. Typically (but not always) the shallowest aquifer at a given location is unconfined, meaning it does not have a confining layer (an aquitard or aquiclude) between it and the surface. The term "perched" refers to ground water accumulating above a low-permeability unit or strata, such as a clay layer. This term is generally used to refer to a small local area of ground water that occurs at an elevation higher than a regionally extensive aquifer. The difference between perched and unconfined aquifers is their size (perched is smaller). Confined aquifers are aquifers that are overlain by a confining layer, often made up of clay. The confining layer might offer some protection from surface contamination.

If the distinction between confined and unconfined is not clear geologically (i.e., if it is not known if a clear confining layer exists, or if the geology is more complex, e.g., a fractured bedrock aquifer), the value of storativity returned from an aquifer test can be used to determine it (although aquifer tests in unconfined aquifers should be interpreted differently than confined ones). Confined aquifers have very low storativity values (much less than 0.01, and as little as 10^{-5}), which means that the aquifer is storing water using the mechanisms of aquifer matrix expansion and the compressibility of water, which typically are both quite small quantities. Unconfined aquifers have storativities (typically then called specific yield) greater than 0.01 (1% of bulk volume); they release water from storage by the mechanism of actually draining the pores of the aquifer, releasing relatively large amounts of water (up to the drainable porosity of the aquifer material, or the minimum volumetric water content).

Isotropic Versus Anisotropic

In isotropic aquifers or aquifer layers the hydraulic conductivity (K) is equal for flow in all directions, while in anisotropic conditions it differs, notably in horizontal (Kh) and vertical (Kv) sense.

Semi-confined aquifers with one or more aquitards work as an anisotropic system, even when the separate layers are isotropic, because the compound Kh and Kv values are different.

When calculating flow to drains or flow to wells in an aquifer, the anisotropy is to be taken into account lest the resulting design of the drainage system may be faulty.

Groundwater in Rock Formations

Map of major US aquifers by rock type

Groundwater may exist in *underground rivers* (e.g., caves where water flows freely underground). This may occur in eroded limestone areas known as karst topography, which make up only a small percentage of Earth's area. More usual is that the pore spaces of rocks in the subsurface are simply saturated with water — like a kitchen sponge — which can be pumped out for agricultural, industrial, or municipal uses.

If a rock unit of low porosity is highly fractured, it can also make a good aquifer (via fissure flow), provided the rock has a hydraulic conductivity sufficient to facilitate movement of water. Porosity is important, but, *alone*, it does not determine a rock's ability to act as an aquifer. Areas of the Deccan Traps (a basaltic lava) in west central India are good examples of rock formations with high porosity but low permeability, which makes them poor aquifers. Similarly, the micro-porous (Upper Cretaceous) Chalk of south east England, although having a reasonably high porosity, has a low grain-to-grain permeability, with its good water-yielding characteristics mostly due to micro-fracturing and fissuring.

Human Dependence on Groundwater

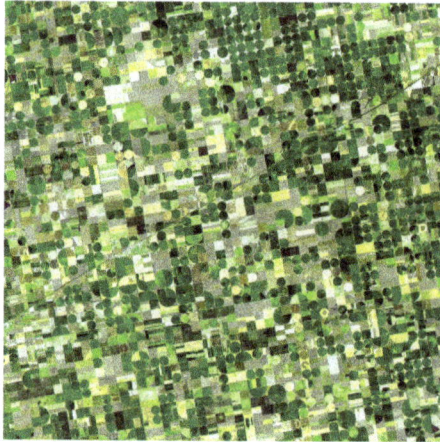

Center-pivot irrigated fields in Kansas covering hundreds of square miles watered by the Ogallala Aquifer

Most land areas on Earth have some form of aquifer underlying them, sometimes at significant depths. In some cases, these aquifers are rapidly being depleted by the human population.

Fresh-water aquifers, especially those with limited recharge by snow or rain, also known as meteoric water, can be over-exploited and depending on the local hydrogeology, may draw in non-potable water or saltwater intrusion from hydraulically connected aquifers or surface water bodies. This can be a serious problem, especially in coastal areas and other areas where aquifer pumping is excessive. In some areas, the ground water can become contaminated by arsenic and other mineral poisons.

Aquifers are critically important in human habitation and agriculture. Deep aquifers in arid areas have long been water sources for irrigation. Many villages and even large cities draw their water supply from wells in aquifers.

Municipal, irrigation, and industrial water supplies are provided through large wells. Multiple wells for one water supply source are termed "wellfields", which may withdraw water from confined or unconfined aquifers. Using ground water from deep, confined aquifers provides more protection from surface water contamination. Some wells, termed "collector wells," are specifically designed to induce infiltration of surface (usually river) water.

Aquifers that provide sustainable fresh groundwater to urban areas and for agricultural irrigation are typically close to the ground surface (within a couple of hundred metres) and have some recharge by fresh water. This recharge is typically from rivers or meteoric water (precipitation) that percolates into the aquifer through overlying unsaturated materials.

Occasionally, sedimentary or "fossil" aquifers are used to provide irrigation and drinking water to urban areas. In Libya, for example, Muammar Gaddafi's Great Manmade River project has pumped large amounts of groundwater from aquifers beneath the Sahara to populous areas near the coast. Though this has saved Libya money over the alternative, desalination, the aquifers are likely to run dry in 60 to 100 years. Aquifer depletion has been cited as one of the causes of the food price rises of 2011.

Subsidence

In unconsolidated aquifers, groundwater is produced from pore spaces between particles of gravel, sand, and silt. If the aquifer is confined by low-permeability layers, the reduced water pressure in the sand and gravel causes slow drainage of water from the adjoining confining layers. If these confining layers are composed of compressible silt or clay, the loss of water to the aquifer reduces the water pressure in the confining layer, causing it to compress from the weight of overlying geologic materials. In severe cases, this compression can be observed on the ground surface as subsidence. Unfortunately, much of the subsidence from groundwater extraction is permanent (elastic rebound is small). Thus, the subsidence is not only permanent, but the compressed aquifer has a permanently reduced capacity to hold water.

Saltwater Intrusion

Aquifers near the coast have a lens of freshwater near the surface and denser seawater under freshwater. Seawater penetrates the aquifer diffusing in from the ocean and is denser than freshwater. For porous (i.e., sandy) aquifers near the coast, the thickness of freshwater atop saltwater is about 40 feet (12 m) for every 1 ft (0.30 m) of freshwater head above sea level. This relationship is called the Ghyben-Herzberg equation. If too much ground water is pumped near the coast, salt-water may intrude into freshwater aquifers causing contamination of potable freshwater supplies. Many coastal aquifers, such as the Biscayne Aquifer near Miami and the New Jersey Coastal Plain aquifer, have problems with saltwater intrusion as a result of overpumping and sea level rise.

Salination

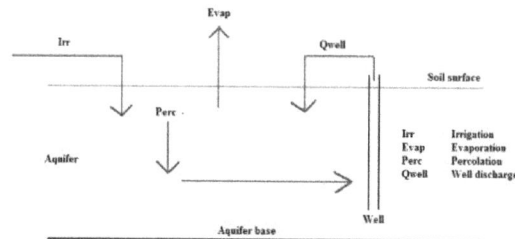

Water balance in the aquifer of a surface irrigated area
with reuse of percolation water by pumping from wells

Diagram of a water balance of the aquifer

Aquifers in surface irrigated areas in semi-arid zones with reuse of the unavoidable irrigation water losses percolating down into the underground by supplemental irrigation from wells run the risk of salination.

Surface irrigation water normally contains salts in the order of 0.5 g/l or more and the annual irrigation requirement is in the order of 10000 m³/ha or more so the annual import of salt is in the order of 5000 kg/ha or more.

Under the influence of continuous evaporation, the salt concentration of the aquifer water may increase continually and eventually cause an environmental problem.

For salinity control in such a case, annually an amount of drainage water is to be discharged from the aquifer by means of a subsurface drainage system and disposed of through a safe outlet. The drainage system may be *horizontal* (i.e. using pipes, tile drains or ditches) or *vertical* (drainage by wells). To estimate the drainage requirement, the use of a groundwater model with an agro-hydro-salinity component may be instrumental, e.g. SahysMod.

Examples

The Great Artesian Basin situated in Australia is arguably the largest groundwater aquifer in the world (over 1.7 million km²). It plays a large part in water supplies for Queensland and remote parts of South Australia.

The Guarani Aquifer, located beneath the surface of Argentina, Brazil, Paraguay, and Uruguay, is one of the world's largest aquifer systems and is an important source of fresh water. Named after the Guarani people, it covers 1,200,000 km², with a volume of about 40,000 km³, a thickness of between 50 m and 800 m and a maximum depth of about 1,800 m.

Aquifer depletion is a problem in some areas, and is especially critical in northern Africa, for example the Great Manmade River project of Libya. However, new methods of groundwater management such as artificial recharge and injection of surface waters during seasonal wet periods has extended the life of many freshwater aquifers, especially in the United States.

The Ogallala Aquifer of the central United States is one of the world's great aquifers, but in places it is being rapidly depleted by growing municipal use, and continuing agricultural use. This huge aquifer, which underlies portions of eight states, contains primarily fossil water from the time of the last glaciation. Annual recharge, in the more arid parts of the aquifer, is estimated to total only about 10 percent of annual withdrawals. According to a 2013 report by research hydrologist Leonard F. Konikow at the United States Geological Survey (USGC), the depletion between 2001–2008, inclusive, is about 32 percent of the cumulative depletion during the entire 20th century (Konikow 2013:22)." In the United States, the biggest users of water from aquifers include agricultural irrigation and oil and coal extraction. "Cumulative total groundwater depletion in the United States accelerated in the late 1940s and continued at an almost steady linear rate through the end of the century. In addition to widely recognized environmental consequences, groundwater depletion also adversely impacts the long-term sustainability of groundwater supplies to help meet the Nation's water needs."

An example of a significant and sustainable carbonate aquifer is the Edwards Aquifer in central Texas. This carbonate aquifer has historically been providing high quality water for nearly 2 million people, and even today, is full because of tremendous recharge from a number of area streams, rivers and lakes. The primary risk to this resource is human development over the recharge areas.

Discontinuous sand bodies at the base of the McMurray Formation in the Athabasca Oil Sands region of northeastern Alberta, Canada, are commonly referred to as the Basal Water Sand (BWS) aquifers. Saturated with water, they are confined beneath impermeable bitumen-saturated sands that are exploited to recover bitumen for synthetic crude oil production. Where they are deep-lying and recharge occurs from underlying Devonian formations they are saline, and where they are shallow and recharged by meteoric water they are non-saline. The BWS typically pose problems for the recovery of bitumen, whether by open-pit mining or by *in situ* methods such as steam-assisted gravity drainage (SAGD), and in some areas they are targets for waste-water injection.

References

- Black, J. T.; Kohser, R. A. (2012). DeGarmo's materials and processes in manufacturing. Wiley. p. 226. ISBN 978-0-470-92467-9.

- Carter, C. B.; Norton, M. G. (2007). Ceramic materials: Science and engineering. Springer. pp. 3 & 4. ISBN 978-0-387-46271-4.

- Carter, C. B.; Norton, M. G. (2007). Ceramic materials: Science and engineering. Springer. pp. 20 & 21. ISBN 978-0-387-46271-4.

- Wachtman, John B., Jr. (ed.) (1999) Ceramic Innovations in the 20th century, The American Ceramic Society. ISBN 978-1-57498-093-6.

- ASTM, E06, "E2121-08 Standard Practice for Installing Radon Mitigation Systems in Existing Low-Rise Residential Buildings", ASTM, 2008, pp. 644-656 ISBN 978-0-8031-5768-2

- L. Mead and G.S. Austin "Dimension Stone", Industrial Minerals and Rocks, 7th Edition, Littleton CO: AIME-Society of Mining Engineers, 2005, pp. 907-923 ISBN 0-87335-233-5

- ASTM, C18, "C1496-01 Standard Guide for Assessment and Maintenance of Exterior Dimension Stone Masonry Walls and Facades", ASTM, 2007, pp. 519-523 ISBN 0-8031-4104-1

8

Allied Fields of Sedimentology

Geology and petrology are considered to be the allied fields of sedimentology. Geology is the study of solid Earth whereas petrology is an important branch of geology that studies the structure of rocks and their origin as well. This chapter will provide a glimpse of related fields of sedimentology briefly.

Geology

Geology is an earth science comprising the study of solid Earth, the rocks of which it is composed, and the processes by which they change. Geology can also refer generally to the study of the solid features of any celestial body (such as the geology of the Moon or Mars).

Geology gives insight into the history of the Earth by providing the primary evidence for plate tectonics, the evolutionary history of life, and past climates. Geology is important for mineral and hydrocarbon exploration and exploitation, evaluating water resources, understanding of natural hazards, the remediation of environmental problems, and for providing insights into past climate change. Geology also plays a role in geotechnical engineering and is a major academic discipline.

Geologic Materials

The majority of geological data comes from research on solid Earth materials. These typically fall into one of two categories: rock and unconsolidated material.

Rock

The majority of research in geology is associated with the study of rock, as rock provides the primary record of the majority of the geologic history of the Earth. There are three major types of rock: igneous, sedimentary, and metamorphic. The rock cycle is an important concept in geology which illustrates the relationships between these three types of rock, and magma. When a rock crystallizes from melt (magma and/or lava), it is an igneous rock. This rock can be weathered and eroded, and then redeposited and lithified into a sedimentary rock, or be turned into a metamorphic rock due to heat and pressure that change the mineral content of the rock which gives it a characteristic fabric. The sedimentary rock can then be subsequently turned into a metamorphic rock due to heat and pressure and is then weathered, eroded, deposited, and lithified, ultimately becoming a sedimentary rock. Sedimentary rock may also be re-eroded and redeposited, and metamorphic rock may also undergo additional metamorphism. All three types of rocks may be re-melted; when this happens, a new magma is formed, from which an igneous rock may once again crystallize.

Rock Cycle

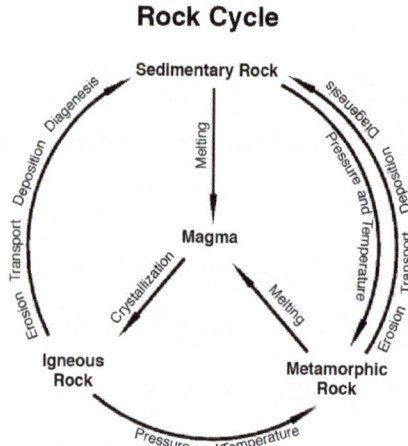

This schematic diagram of the rock cycle shows the relationship between magma and sedimentary, metamorphic, and igneous rock

Unconsolidated Material

Geologists also study unlithified material, which typically comes from more recent deposits. These materials are superficial deposits which lie above the bedrock. Because of this, the study of such material is often known as Quaternary geology, after the recent Quaternary Period. This includes the study of sediment and soils, including studies in geomorphology, sedimentology, and paleoclimatology.

Whole-Earth Structure

Plate Tectonics

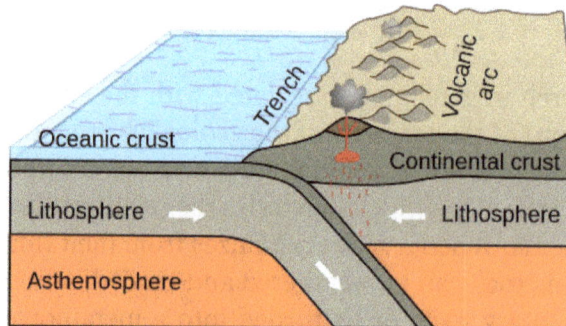

Oceanic-continental convergence resulting in subduction and volcanic arcs illustrates one effect of plate tectonics.

In the 1960s, it was discovered that the Earth's lithosphere, including the crust and rigid uppermost portion of the upper mantle, is separated into tectonic plates that move across the plastically deforming, solid, upper mantle, which is called the astheno-sphere. This theory is supported by a several types of observations, including seafloor spreading, and the global distribution of mountain terrain and seismicity.

There is an intimate coupling between the movement of the plates on the surface and the convection of the mantle: oceanic plate motions and mantle convection currents always move in the same direction, because the oceanic lithosphere is the rigid upper thermal boundary layer of the convecting mantle. This coupling be-tween rigid plates moving on the surface of the Earth and the convecting mantle is called plate tectonics.

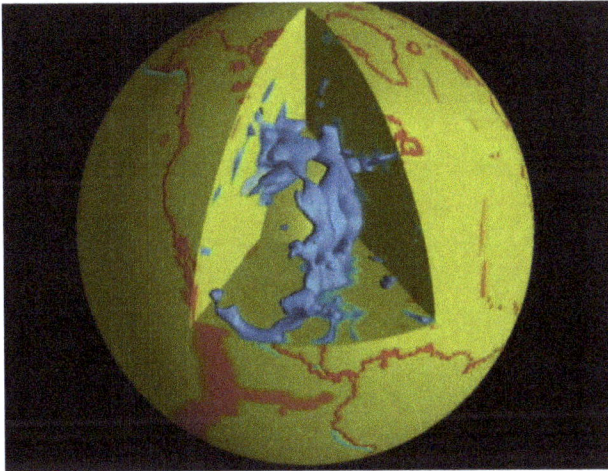

On this diagram, subducting slabs are in blue, and continental margins and a few plate boundaries are in red. The blue blob in the cutaway section is the seismically imaged Farallon Plate, which is subducting beneath North America. The remnants of this plate on the Surface of the Earth are the Juan de Fuca Plate and Explorer plate in the Northwestern USA / Southwestern Canada, and the Cocos Plate on the west coast of Mexico.

The development of plate tectonics provided a physical basis for many observations of the solid Earth. Long linear regions of geologic features could be explained as plate boundaries. Mid-ocean ridges, high regions on the seafloor where hydrothermal vents and volcanoes exist, were explained as divergent boundaries, where two plates move apart. Arcs of volcanoes and earthquakes were explained as convergent boundaries, where one plate subducts under another. Transform boundaries, such as the San An-dreas Fault system, resulted in widespread powerful earthquakes. Plate tectonics also provided a mechanism for Alfred Wegener's theory of continental drift, in which the continents move across the surface of the Earth over geologic time. They also provided a driving force for crustal deformation, and a new setting for the observations of struc-tural geology. The power of the theory of plate tectonics lies in its ability to combine all of these observations into a single theory of how the lithosphere moves over the convecting mantle.

Earth Structure

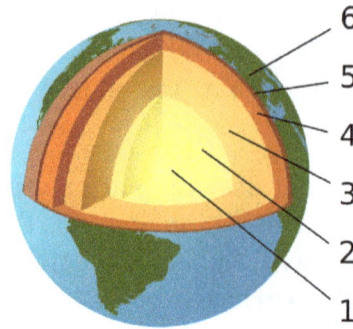

The Earth's layered structure. (1) inner core; (2) outer core; (3) lower mantle; (4) upper mantle; (5) lithosphere; (6) crust (part of the lithosphere)

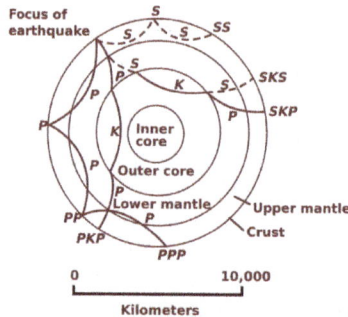

Earth layered structure. Typical wave paths from earthquakes like these gave early seismologists insights into the layered structure of the Earth

Advances in seismology, computer modeling, and mineralogy and crystallography at high temperatures and pressures give insights into the internal composition and structure of the Earth.

Seismologists can use the arrival times of seismic waves in reverse to image the interior of the Earth. Early advances in this field showed the existence of a liquid outer core (where shear waves were not able to propagate) and a dense solid inner core. These advances led to the development of a layered model of the Earth, with a crust and lithosphere on top, the mantle below (separated within itself by seismic discontinuities at 410 and 660 kilometers), and the outer core and inner core below that. More recently, seismologists have been able to create detailed images of wave speeds inside the earth in the same way a doctor images a body in a CT scan. These images have led to a much more detailed view of the interior of the Earth, and have replaced the simplified layered model with a much more dynamic model.

Mineralogists have been able to use the pressure and temperature data from the seismic and modelling studies alongside knowledge of the elemental composition of the Earth to reproduce these conditions in experimental settings and measure changes in crystal structure. These studies explain the chemical changes associated with the major seismic discontinuities in the mantle and show the crystallographic structures expected in the inner core of the Earth.

Geologic Time

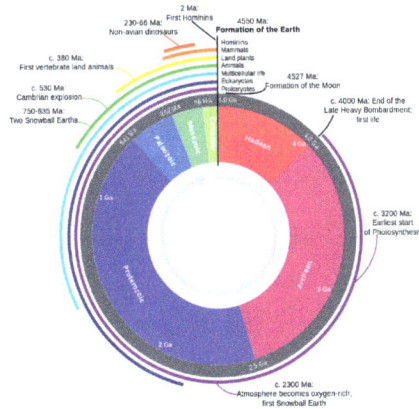

Geological time put in a diagram called a geological clock, showing the relative lengths of the eons of the Earth's history.

The geologic time scale encompasses the history of the Earth. It is bracketed at the old end by the dates of the earliest Solar System material at 4.567 Ga, (gigaannum: billion years ago) and the age of the Earth at 4.54 Ga at the beginning of the informally recognized Hadean eon. At the young end of the scale, it is bracketed by the present day in the Holocene epoch.

Important Milestones

- 4.567 Ga: Solar system formation

- 4.54 Ga: Accretion of Earth

- c. 4 Ga: End of Late Heavy Bombardment, first life

- c. 3.5 Ga: Start of photosynthesis

- c. 2.3 Ga: Oxygenated atmosphere, first snowball Earth

- 730–635 Ma (megaannum: million years ago): second snowball Earth

- 542 ± 0.3 Ma: Cambrian explosion – vast multiplication of hard-bodied life; first abundant fossils; start of the Paleozoic

- c. 380 Ma: First vertebrate land animals

- 250 Ma: Permian-Triassic extinction – 90% of all land animals die; end of Paleozoic and beginning of Mesozoic

- 66 Ma: Cretaceous–Paleogene extinction – Dinosaurs die; end of Mesozoic and beginning of Cenozoic

- c. 7 Ma: First hominins appear

- 3.9 Ma: First Australopithecus, direct ancestor to modern Homo sapiens, appear

- 200 ka (kiloannum: thousand years ago): First modern Homo sapiens appear in East Africa

Brief Time Scale

The following four timelines show the geologic time scale. The first shows the entire time from the formation of the Earth to the present, but this compresses the most recent eon. Therefore, the second scale shows the most recent eon with an expanded scale. The second scale compresses the most recent era, so the most recent era is expanded in the third scale. The third scale compresses the most recent period, so the most recent period is expanded in the fourth scale.

Dating Methods

Relative Dating

Methods for relative dating were developed when geology first emerged as a formal science. Geologists still use the following principles today as a means to provide information about geologic history and the timing of geologic events.

The principle of Uniformitarianism states that the geologic processes observed in operation that modify the Earth's crust at present have worked in much the same way over geologic time. A fundamental principle of geology advanced by the 18th century Scottish physician and geologist James Hutton is that "the present is the key to the past." In Hutton's words: "the past history of our globe must be explained by what can be seen to be happening now."

The principle of intrusive relationships concerns crosscutting intrusions. In geology, when an igneous intrusion cuts across a formation of sedimentary rock, it can be determined that the igneous intrusion is younger than the sedimentary rock. Different types of intrusions include stocks, laccoliths, batholiths, sills and dikes.

The principle of cross-cutting relationships pertains to the formation of faults and the age of the sequences through which they cut. Faults are younger than the rocks they cut; accordingly, if a fault is found that penetrates some formations but not those on top of it, then the formations that were cut are older than the fault, and the ones that are not cut must be younger than the fault. Finding the key bed in these situations may help determine whether the fault is a normal fault or a thrust fault.

The principle of inclusions and components states that, with sedimentary rocks, if in-clusions (or *clasts*) are found in a formation, then the inclusions must be older than the formation that contains them. For example, in sedimentary rocks, it is common for gravel from an older formation to be ripped up and included in a newer layer. A similar situation with igneous rocks occurs when xenoliths are found. These foreign bodies are picked up as magma or lava flows, and are incorporated, later to cool in the matrix. As a result, xenoliths are older than the rock which contains them.

The Permian through Jurassic stratigraphy of the Colorado Plateau area of southeastern Utah is an example of both original horizontality and the law of superposition. These strata make up much of the famous prominent rock formations in widely spaced protect-ed areas such as Capitol Reef National Park and Canyonlands National Park. From top to bottom: Rounded tan domes of the Navajo Sandstone, layered red Kayenta Formation, cliff-forming, vertically jointed, red Wingate Sandstone, slope-forming, purplish Chinle Formation, layered, lighter-red Moenkopi Formation, and white, layered Cutler Forma-tion sandstone. Picture from Glen Canyon National Recreation Area, Utah.

The principle of original horizontality states that the deposition of sediments occurs as essentially horizontal beds. Observation of modern marine and non-marine sediments in a wide variety of environments supports this generalization (although cross-bedding is inclined, the overall orientation of cross-bedded units is horizontal).

The principle of superposition states that a sedimentary rock layer in a tectonically undisturbed sequence is younger than the one beneath it and older than the one above it. Logically a younger layer cannot slip beneath a layer previously deposited. This prin-ciple allows sedimentary layers to be viewed as a form of vertical time line, a partial or complete record of the time elapsed from deposition of the lowest layer to deposition of the highest bed.

The principle of faunal succession is based on the appearance of fossils in sedimentary rocks. As organisms exist at the same time period throughout the world, their pres-ence or (sometimes) absence may be used to provide a relative age of the formations in which they are found. Based on principles laid out by William Smith almost a hundred years before the publication of Charles Darwin's theory of evolution, the principles of succession were developed independently of evolutionary thought. The principle be-comes quite complex, however, given the uncertainties of fossilization, the localization of fossil types due to lateral changes in habitat (facies change in sedimentary strata), and that not all fossils may be found globally at the same time.

Absolute Dating

Geologists also use methods to determine the absolute age of rock samples and geolog-ical events. These dates are useful on their own and may also be used in conjunction with relative dating methods or to calibrate relative methods.

At the beginning of the 20th century, advancement in geological science was facilitated by the ability to obtain accurate absolute dates to geologic events using radioactive isotopes and other methods. This changed the understanding of geologic time. Previously, geologists could only use fossils and stratigraphic correlation to date sections of rock relative to one another. With isotopic dates, it became possible to assign absolute ages to rock units, and these absolute dates could be applied to fossil sequences in which there was datable material, converting the old relative ages into new absolute ages.

For many geologic applications, isotope ratios of radioactive elements are measured in minerals that give the amount of time that has passed since a rock passed through its particular closure temperature, the point at which different radiometric isotopes stop diffusing into and out of the crystal lattice. These are used in geochronologic and thermochronologic studies. Common methods include uranium-lead dating, potassium-argon dating, argon-argon dating and uranium-thorium dating. These methods are used for a variety of applications. Dating of lava and volcanic ash layers found within a stratigraphic sequence can provide absolute age data for sedimentary rock units which do not contain radioactive isotopes and calibrate relative dating techniques. These methods can also be used to determine ages of pluton emplacement. Thermochemical techniques can be used to determine temperature profiles within the crust, the uplift of mountain ranges, and paleotopography.

Fractionation of the lanthanide series elements is used to compute ages since rocks were removed from the mantle.

Other methods are used for more recent events. Optically stimulated luminescence and cosmogenic radionuclide dating are used to date surfaces and/or erosion rates. Dendrochronology can also be used for the dating of landscapes. Radiocarbon dating is used for geologically young materials containing organic carbon.

Geological Development of an Area

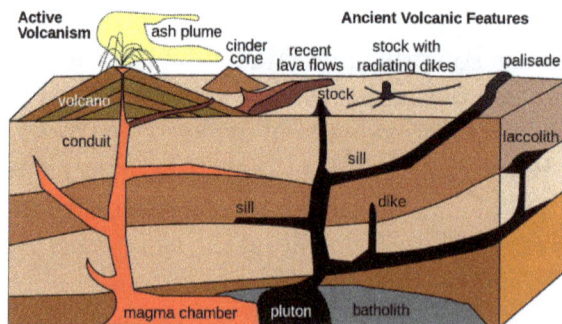

An originally horizontal sequence of sedimentary rocks (in shades of tan) are affected by igneous activity. Deep below the surface are a magma chamber and large associated igneous bodies. The magma chamber feeds the volcano, and sends offshoots of magma that will later crystallize into dikes and sills. Magma also advances upwards to form intrusive igneous bodies. The diagram illustrates both a cinder cone volcano, which releases ash, and a composite volcano, which releases both lava and ash.

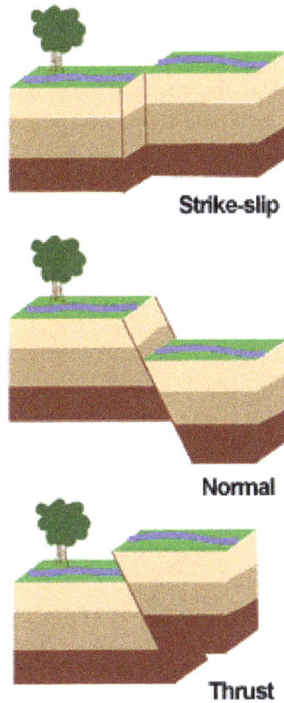

An illustration of the three types of faults. Strike-slip faults occur when rock units slide past one another, normal faults occur when rocks are undergoing horizontal extension, and thrust faults occur when rocks are undergoing horizontal shortening.

The geology of an area changes through time as rock units are deposited and inserted, and deformational processes change their shapes and locations.

Rock units are first emplaced either by deposition onto the surface or intrusion into the overlying rock. Deposition can occur when sediments settle onto the surface of the Earth and later lithify into sedimentary rock, or when as volcanic material such as volcanic ash or lava flows blanket the surface. Igneous intrusions such as batholiths, laccoliths, dikes, and sills, push upwards into the overlying rock, and crystallize as they intrude.

After the initial sequence of rocks has been deposited, the rock units can be deformed and/or metamorphosed. Deformation typically occurs as a result of horizontal shortening, horizontal extension, or side-to-side (strike-slip) motion. These structural regimes broadly relate to convergent boundaries, divergent boundaries, and transform boundaries, respectively, between tectonic plates.

When rock units are placed under horizontal compression, they shorten and become thicker. Because rock units, other than muds, do not significantly change in volume, this is accomplished in two primary ways: through faulting and folding. In the shallow crust, where brittle deformation can occur, thrust faults form, which causes deeper rock to move on top of shallower rock. Because deeper rock is often older, as noted by the principle of superposition, this can result in older rocks moving on top of younger ones. Movement along faults can result in folding, either because the faults are not planar or because rock layers are dragged along, forming drag folds as slip occurs along the fault. Deeper in the Earth, rocks behave plastically and fold instead of faulting. These folds can either be those where the material in the center of the fold buckles upwards, creating "antiforms", or where it buckles downwards, creating "synforms". If the tops of the rock units within the folds remain pointing upwards, they are called anticlines and synclines, respectively. If some of the units in the fold are facing downward, the structure is called an overturned anticline or syncline, and if all of the rock units are overturned or the correct up-direction is unknown, they are simply called by the most general terms, antiforms and synforms.

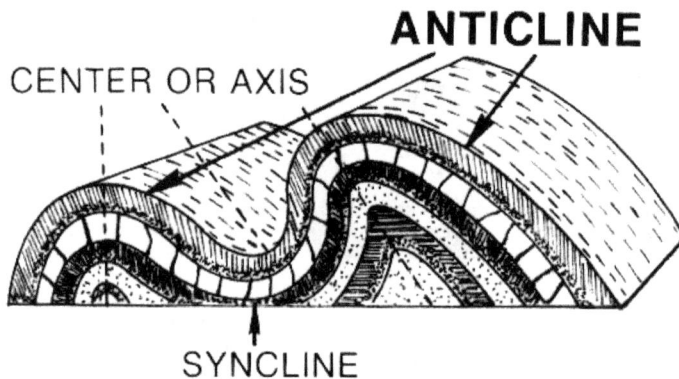

A diagram of folds, indicating an anticline and a syncline.

Even higher pressures and temperatures during horizontal shortening can cause both folding and metamorphism of the rocks. This metamorphism causes changes in the mineral composition of the rocks; creates a foliation, or planar surface, that is related to mineral growth under stress. This can remove signs of the original textures of the rocks, such as bedding in sedimentary rocks, flow features of lavas, and crystal patterns in crystalline rocks.

Extension causes the rock units as a whole to become longer and thinner. This is primarily accomplished through normal faulting and through the ductile stretching and thinning. Normal faults drop rock units that are higher below those that are lower. This typically results in younger units being placed below older units. Stretching of units can result in their thinning; in fact, there is a location within the Maria Fold and Thrust Belt in which the entire sedimentary sequence of the Grand Canyon can be seen over a length of less than a meter. Rocks at the depth to be ductilely stretched are often also metamorphosed. These stretched rocks can also pinch into lenses, known as boudins, after the French word for "sausage", because of their visual similarity.

Where rock units slide past one another, strike-slip faults develop in shallow regions, and become shear zones at deeper depths where the rocks deform ductilely.

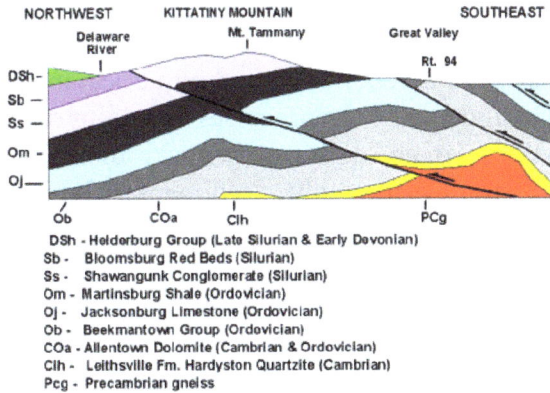

Geologic cross section of Kittatinny Mountain. This cross section shows metamorphic rocks, overlain by younger sediments deposited after the metamorphic event. These rock units were later folded and faulted during the uplift of the mountain.

The addition of new rock units, both depositionally and intrusively, often occurs during deformation. Faulting and other deformational processes result in the creation of topographic gradients, causing material on the rock unit that is increasing in elevation to be eroded by hillslopes and channels. These sediments are deposited on the rock unit that is going down. Continual motion along the fault maintains the topographic gradient in spite of the movement of sediment, and continues to create accommodation space for the material to deposit. Deformational events are often also associated with volcanism and igneous activity. Volcanic ashes and lavas accumulate on the surface, and igneous intrusions enter from below. Dikes, long, planar igneous intrusions, enter along cracks, and therefore often form in large numbers in areas that are being actively deformed. This can result in the emplacement of dike swarms, such as those that are observable across the Canadian shield, or rings of dikes around the lava tube of a volcano.

All of these processes do not necessarily occur in a single environment, and do not necessarily occur in a single order. The Hawaiian Islands, for example, consist almost entirely of layered basaltic lava flows. The sedimentary sequences of the mid-continental United States and the Grand Canyon in the southwestern United States contain almost-undeformed stacks of sedimentary rocks that have remained in place since Cambrian time. Other areas are much more geologically complex. In the southwestern United States, sedimentary, volcanic, and intrusive rocks have been metamorphosed, faulted, foliated, and folded. Even older rocks, such as the Acasta gneiss of the Slave craton in northwestern Canada, the oldest known rock in the world have been metamorphosed to the point where their origin is undiscernable without laboratory analysis. In addition, these processes can occur in stages. In many places, the Grand Canyon in the southwestern United States being a very visible example, the lower rock units were metamorphosed and deformed, and then deformation ended and the upper, undeformed units were deposited. Although any amount of rock emplacement and rock deformation can occur, and they can occur any number of times, these concepts provide a guide to understanding the geological history of an area.

Methods of Geology

Geologists use a number of field, laboratory, and numerical modeling methods to decipher Earth history and to understand the processes that occur on and inside the Earth. In typical geological investigations, geologists use primary information related to petrology (the study of rocks), stratigraphy (the study of sedimentary layers), and structural geology (the study of positions of rock units and their deformation). In many cases, geologists also study modern soils, rivers, landscapes, and glaciers; investigate past and current life and biogeochemical pathways, and use geophysical methods to investigate the subsurface. Sub-specialities of geology may distinguish endogenous and exogenous geology.

Field Methods

A standard Brunton Pocket Transit, commonly used by geologists for mapping and surveying.

A typical USGS field mapping camp in the 1950s

Today, handheld computers with GPS and geographic information systems software are often used in geological field work (digital geologic mapping).

Geological field work varies depending on the task at hand. Typical fieldwork could consist of:

- Geological mapping

    o Structural mapping: identifying the locations of major rock units and the faults and folds that led to their placement there.

    o Stratigraphic mapping: pinpointing the locations of sedimentary facies (lithofacies and biofacies) or the mapping of isopachs of equal thickness of sedimentary rock

    o Surficial mapping: recording the locations of soils and surficial deposits

- Surveying of topographic features

    o compilation of topographic maps

    o Work to understand change across landscapes, including:

        ▪ Patterns of erosion and deposition

        ▪ River-channel change through migration and avulsion

        ▪ Hillslope processes

- Subsurface mapping through geophysical methods

    o These methods include:

        ▪ Shallow seismic surveys

        ▪ Ground-penetrating radar

        ▪ Aeromagnetic surveys

        ▪ Electrical resistivity tomography

    o They aid in:

        ▪ Hydrocarbon exploration

        ▪ Finding groundwater

        ▪ Locating buried archaeological artifacts

- High-resolution stratigraphy

    o Measuring and describing stratigraphic sections on the surface

    o Well drilling and logging

- Biogeochemistry and geomicrobiology

  o Collecting samples to:

    ▪ determine biochemical pathways

    ▪ identify new species of organisms

    ▪ identify new chemical compounds

  o and to use these discoveries to:

    ▪ understand early life on Earth and how it functioned and metabolized

    ▪ find important compounds for use in pharmaceuticals

- Paleontology: excavation of fossil material

  o For research into past life and evolution

  o For museums and education

- Collection of samples for geochronology and thermochronology

- Glaciology: measurement of characteristics of glaciers and their motion

Petrology

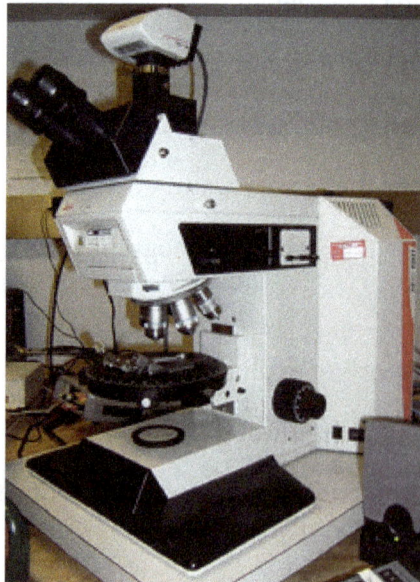

A petrographic microscope - an optical microscope fitted with cross-polarizing lenses, a conoscopic lens, and compensators (plates of anisotropic materials; gypsum plates and quartz wedges are common), for crystallographic analysis.

In addition to identifying rocks in the field (lithology), petrologists identify rock samples in the laboratory. Two of the primary methods for identifying rocks in the laboratory are through optical microscopy and by using an electron microprobe. In an optical mineralogy analysis, petrologists analyze thin sections of rock samples using a petrographic microscope, where the minerals can be identified through their different properties in plane-polarized and cross-polarized light, including their birefringence, pleochroism, twinning, and interference properties with a conoscopic lens. In the electron microprobe, individual locations are analyzed for their exact chemical compositions and variation in composition within individual crystals. Stable and radioactive isotope studies provide insight into the geochemical evolution of rock units.

Petrologists can also use fluid inclusion data and perform high temperature and pressure physical experiments to understand the temperatures and pressures at which different mineral phases appear, and how they change through igneous and metamorphic processes. This research can be extrapolated to the field to understand metamorphic processes and the conditions of crystallization of igneous rocks. This work can also help to explain processes that occur within the Earth, such as subduction and magma chamber evolution.

Structural Geology

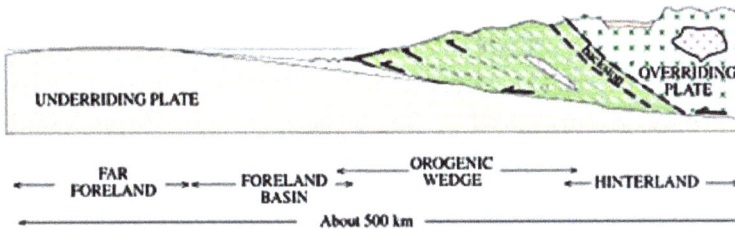

A diagram of an orogenic wedge. The wedge grows through faulting in the interior and along the main basal fault, called the décollement. It builds its shape into a critical taper, in which the angles within the wedge remain the same as failures inside the material balance failures along the décollement. It is analogous to a bulldozer pushing a pile of dirt, where the bulldozer is the overriding plate.

Structural geologists use microscopic analysis of oriented thin sections of geologic samples to observe the fabric within the rocks which gives information about strain within the crystalline structure of the rocks. They also plot and combine measurements of geological structures to better understand the orientations of faults and folds to reconstruct the history of rock deformation in the area. In addition, they perform analog and numerical experiments of rock deformation in large and small settings.

The analysis of structures is often accomplished by plotting the orientations of various features onto stereonets. A stereonet is a stereographic projection of a sphere onto a plane, in which planes are projected as lines and lines are projected as points. These can be used to find the locations of fold axes, relationships between faults, and relationships between other geologic structures.

Among the most well-known experiments in structural geology are those involving orogenic wedges, which are zones in which mountains are built along convergent tectonic plate boundaries. In the analog versions of these experiments, horizontal layers of sand are pulled along a lower surface into a back stop, which results in realistic-looking patterns of faulting and the growth of a critically tapered (all angles remain the same) orogenic wedge. Numerical models work in the same way as these analog models, though they are often more sophisticated and can include patterns of erosion and uplift in the mountain belt. This helps to show the relationship between erosion and the shape of a mountain range. These studies can also give useful information about pathways for metamorphism through pressure, temperature, space, and time.

Stratigraphy

In the laboratory, stratigraphers analyze samples of stratigraphic sections that can be returned from the field, such as those from drill cores. Stratigraphers also analyze data from geophysical surveys that show the locations of stratigraphic units in the subsurface. Geophysical data and well logs can be combined to produce a better view of the subsurface, and stratigraphers often use computer programs to do this in three dimensions. Stratigraphers can then use these data to reconstruct ancient processes occurring on the surface of the Earth, interpret past environments, and locate areas for water, coal, and hydrocarbon extraction.

In the laboratory, biostratigraphers analyze rock samples from outcrop and drill cores for the fossils found in them. These fossils help scientists to date the core and to understand the depositional environment in which the rock units formed. Geochronologists precisely date rocks within the stratigraphic section to provide better absolute bounds on the timing and rates of deposition. Magnetic stratigraphers look for signs of magnetic reversals in igneous rock units within the drill cores. Other scientists perform stable-isotope studies on the rocks to gain information about past climate.

Planetary Geology

Surface of Mars as photographed by the Viking 2 lander December 9, 1977.

With the advent of space exploration in the twentieth century, geologists have begun to look at other planetary bodies in the same ways that have been developed to study the Earth. This new field of study is called planetary geology (sometimes known as astrogeology) and relies on known geologic principles to study other bodies of the solar system.

Although the Greek-language-origin prefix *geo* refers to Earth, "geology" is often used in conjunction with the names of other planetary bodies when describing their composition and internal processes: examples are "the geology of Mars" and "Lunar geology". Specialised terms such as *selenology* (studies of the Moon), *areology* (of Mars), etc., are also in use.

Although planetary geologists are interested in studying all aspects of other planets, a significant focus is to search for evidence of past or present life on other worlds. This has led to many missions whose primary or ancillary purpose is to examine planetary bodies for evidence of life. One of these is the Phoenix lander, which analyzed Martian polar soil for water, chemical, and mineralogical constituents related to biological processes.

Applied Geology

Economic Geology

Economic geology is an important branch of geology which deals with different aspects of economic minerals being used by humankind to fulfill its various needs. The economic minerals are those which can be extracted profitably. Economic geologists help locate and manage the Earth's natural resources, such as petroleum and coal, as well as mineral resources, which include metals such as iron, copper, and uranium.

Mining Geology

Mining geology consists of the extractions of mineral resources from the Earth. Some resources of economic interests include gemstones, metals, and many minerals such as asbestos, perlite, mica, phosphates, zeolites, clay, pumice, quartz, and silica, as well as elements such as sulfur, chlorine, and helium.

Petroleum Geology

Petroleum geologists study the locations of the subsurface of the Earth which can contain extractable hydrocarbons, especially petroleum and natural gas. Because many of these reservoirs are found in sedimentary basins, they study the formation of these basins, as well as their sedimentary and tectonic evolution and the present-day positions of the rock units.

Engineering Geology

Engineering geology is the application of the geologic principles to engineering practice for the purpose of assuring that the geologic factors affecting the location, design, construction, operation, and maintenance of engineering works are properly addressed.

In the field of civil engineering, geological principles and analyses are used in order to ascertain the mechanical principles of the material on which structures are built. This allows tunnels to be built without collapsing, bridges and skyscrapers to be built with sturdy foundations, and buildings to be built that will not settle in clay and mud.

Hydrology and Environmental Issues

Geology and geologic principles can be applied to various environmental problems such as stream restoration, the restoration of brownfields, and the understanding of the interaction between natural habitat and the geologic environment. Groundwater hydrology, or hydrogeology, is used to locate groundwater, which can often provide a ready supply of uncontaminated water and is especially important in arid regions, and to monitor the spread of contaminants in groundwater wells.

Geologists also obtain data through stratigraphy, boreholes, core samples, and ice cores. Ice cores and sediment cores are used to for paleoclimate reconstructions, which tell geologists about past and present temperature, precipitation, and sea level across the globe. These datasets are our primary source of information on global climate change outside of instrumental data.

Natural Hazards

Geologists and geophysicists study natural hazards in order to enact safe building codes and warning systems that are used to prevent loss of property and life. Examples of important natural hazards that are pertinent to geology (as opposed those that are mainly or only pertinent to meteorology) are:

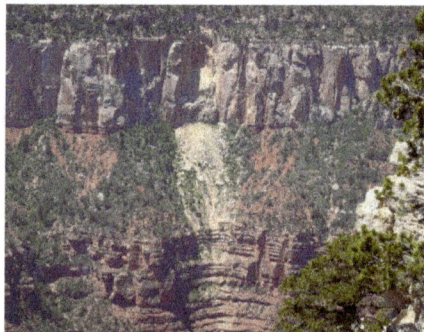

Rockfall in the Grand Canyon

- Avalanches

- Earthquakes

- Floods

- Landslides and debris flows

- River channel migration and avulsion

- Liquefaction

- Sinkholes

- Subsidence

- Tsunamis

- Volcanoes

History of Geology

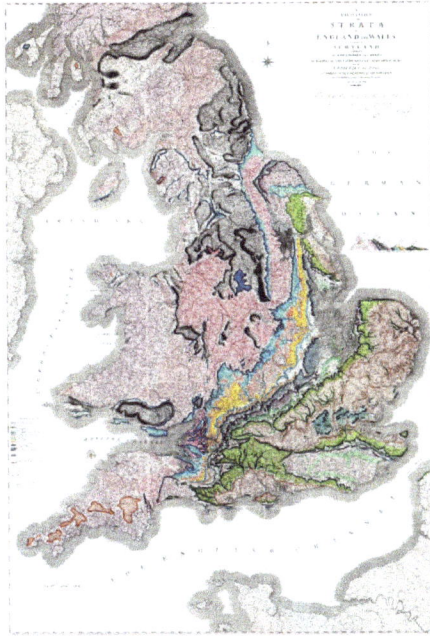

William Smith's geologic map of England, Wales, and southern Scotland. Completed in 1815, it was the second national-scale geologic map, and by far the most accurate of its time.

The study of the physical material of the Earth dates back at least to ancient Greece when Theophrastus (372–287 BCE) wrote the work *Peri Lithon* (*On Stones*). During the Roman period, Pliny the Elder wrote in detail of the many minerals and metals then in practical use – even correctly noting the origin of amber.

Some modern scholars, such as Fielding H. Garrison, are of the opinion that the origin of the science of geology can be traced to Persia after the Muslim conquests had come to an end. Abu al-Rayhan al-Biruni (973–1048 CE) was one of the earliest Persian geologists, whose works included the earliest writings on the geology of India, hypothesizing that the Indian subcontinent was once a sea. Drawing from Greek and Indian scientific literature that were not destroyed by the Muslim conquests, the Persian scholar Ibn Sina (Avicenna, 981–1037) proposed detailed explanations for the formation of mountains, the origin of earthquakes, and other topics central to modern geology, which provided an essential foundation for the later development of the science. In China, the polymath Shen Kuo (1031–1095) formulated a hypothesis for the process of land formation: based on his observation of fossil animal shells in a geological stratum in a mountain hundreds of miles from the ocean, he inferred that the land was formed by erosion of the mountains and by deposition of silt.

Nicolas Steno (1638–1686) is credited with the law of superposition, the principle of original horizontality, and the principle of lateral continuity: three defining principles of stratigraphy.

The word *geology* was first used by Ulisse Aldrovandi in 1603, then by Jean-André Deluc in 1778 and introduced as a fixed term by Horace-Bénédict de Saussure in 1779. The word is derived from the *logos*, meaning "speech". But according to another source, the word "geology" comes from a Norwegian, Mikkel Pedersøn Escholt (1600–1699), who was a priest and scholar. Escholt first used the definition in his book titled, *Geologica Norvegica* (1657).

William Smith (1769–1839) drew some of the first geological maps and began the process of ordering rock strata (layers) by examining the fossils contained in them.

James Hutton is often viewed as the first modern geologist. In 1785 he presented a paper entitled *Theory of the Earth* to the Royal Society of Edinburgh. In his paper, he explained his theory that the Earth must be much older than had previously been supposed to allow enough time for mountains to be eroded and for sediments to form new rocks at the bottom of the sea, which in turn were raised up to become dry land. Hutton published a two-volume version of his ideas in 1795 (Vol. 1, Vol. 2).

Scotsman James Hutton, father of modern geology

Followers of Hutton were known as *Plutonists* because they believed that some rocks were formed by *vulcanism*, which is the deposition of lava from volcanoes, as opposed to the *Neptunists*, led by Abraham Werner, who believed that all rocks had settled out of a large ocean whose level gradually dropped over time.

The first geological map of the U.S. was produced in 1809 by William Maclure. In 1807, Maclure commenced the self-imposed task of making a geological survey of the United States. Almost every state in the Union was traversed and mapped by him, the Allegheny Mountains being crossed and recrossed some 50 times. The results of his unaided labours were submitted to the American Philosophical Society in a memoir entitled *Observations on the Geology of the United States explanatory of a Geological Map*, and published in the *Society's Transactions*, together with the nation's first geological map. This antedates William Smith's geological map of England by six years, although it was constructed using a different classification of rocks.

Sir Charles Lyell first published his famous book, *Principles of Geology*, in 1830. This book, which influenced the thought of Charles Darwin, successfully promoted the doctrine of uniformitarianism. This theory states that slow geological processes have occurred throughout the Earth's history and are still occurring today. In contrast, catastrophism is the theory that Earth's features formed in single, catastrophic events and remained unchanged thereafter. Though Hutton believed in uniformitarianism, the idea was not widely accepted at the time.

Much of 19th-century geology revolved around the question of the Earth's exact age. Estimates varied from a few hundred thousand to billions of years. By the early 20th century, radiometric dating allowed the Earth's age to be estimated at two billion years. The awareness of this vast amount of time opened the door to new theories about the processes that shaped the planet.

Some of the most significant advances in 20th-century geology have been the development of the theory of plate tectonics in the 1960s and the refinement of estimates of the planet's age. Plate tectonics theory arose from two separate geological observations: seafloor spreading and continental drift. The theory revolutionized the Earth sciences. Today the Earth is known to be approximately 4.5 billion years old.

Petrology

Petrology is the branch of geology that studies the origin, composition, distribution and structure of rocks.

Lithology was once approximately synonymous with petrography, but in current usage, lithology focuses on macroscopic hand-sample or outcrop-scale description of rocks while petrography is the speciality that deals with microscopic details.

In the petroleum industry, lithology, or more specifically mud logging, is the graphic representation of geological formations being drilled through, and drawn on a log called a mud log. As the cuttings are circulated out of the borehole they are sampled, examined (typically under a 10× microscope) and tested chemically when needed.

Methodology

Petrology utilizes the fields of mineralogy, petrography, optical mineralogy, and chemical analysis to describe the composition and texture of rocks. Petrologists also include the principles of geochemistry and geophysics through the study of geochemical trends and cycles and the use of thermodynamic data and experiments in order to better understand the origins of rocks.

Branches

There are three branches of petrology, corresponding to the three types of rocks: igneous, metamorphic, and sedimentary, and another dealing with experimental techniques:

- Igneous petrology focuses on the composition and texture of igneous rocks (rocks such as granite or basalt which have crystallized from molten rock or magma). Igneous rocks include volcanic and plutonic rocks.

- Sedimentary petrology focuses on the composition and texture of sedimentary rocks (rocks such as sandstone, shale, or limestone which consist of pieces or particles derived from other rocks or biological or chemical deposits, and are usually bound together in a matrix of finer material).

- Metamorphic petrology focuses on the composition and texture of metamorphic rocks (rocks such as slate, marble, gneiss, or schist which started out as sedimentary or igneous rocks but which have undergone chemical, mineralogical or textural changes due to extremes of pressure, temperature or both)

- Experimental petrology employs high-pressure, high-temperature apparatus to investigate the geochemistry and phase relations of natural or synthetic materials at elevated pressures and temperatures. Experiments are particularly useful for investigating rocks of the lower crust and upper mantle that rarely survive the journey to the surface in pristine condition. They are also one of the prime sources of information about completely inaccessible rocks such as those in the Earth's lower mantle and in the mantles of the other terrestrial planets and the Moon. The work of experimental petrologists has laid a foundation on which modern understanding of igneous and metamorphic processes has been built.

References

- Dalrymple, G. Brent (1994). The age of the earth. Stanford, California: Stanford Univ. Press. ISBN 0-8047-2331-1.

- Levin, Harold L. (2010). The earth through time (9th ed.). Hoboken, N.J.: J. Wiley. p. 18. ISBN 978-0-470-38774-0.

- Faure, Gunter (1998). Principles and applications of geochemistry: a comprehensive textbook for geology students. Upper Saddle River, NJ: Prentice-Hall. ISBN 978-0-02-336450-1.

- Burger, H. Robert; Sheehan, Anne F.; Jones, Craig H. (2006). Introduction to applied geophysics : exploring the shallow subsurface. New York: W.W. Norton. ISBN 0-393-92637-0.

- Krumbein, Wolfgang E., ed. (1978). Environmental biogeochemistry and geomicrobiology. Ann Arbor, Mich.: Ann Arbor Science Publ. ISBN 0-250-40218-1.

- Hubbard, Bryn; Glasser, Neil (2005). Field techniques in glaciology and glacial geomorphology. Chichester, England: J. Wiley. ISBN 0-470-84426-4.

- Nesse, William D. (1991). Introduction to optical mineralogy. New York: Oxford University Press. ISBN 0-19-506024-5.

- Shepherd, T.J.; Rankin, A.H.; Alderton, D.H.M. (1985). A practical guide to fluid inclusion studies. Glasgow: Blackie. ISBN 0-412-00601-4.

- McBirney, Alexander R. (2007). Igneous petrology. Boston: Jones and Bartlett Publishers. ISBN 978-0-7637-3448-0.

- Spear, Frank S. (1995). Metamorphic phase equilibria and pressure-temperature-time paths. Washington, DC: Mineralogical Soc. of America. ISBN 978-0-939950-34-8.

- Bally, A.W., ed. (1987). Atlas of seismic stratigraphy. Tulsa, Okla., U.S.A.: American Association of Petroleum Geologists. ISBN 0-89181-033-1.

Permissions

Index

A

Absolute Dating, 265
Aeolian Processes, 7, 12, 22, 25, 69, 127
Algoma Type, 100
Alluvial Fan, 30, 39-40, 109, 196
Antidunes, 11, 20, 111-112, 117
Aquifer, 219, 250-257

B

Banded Iron Formations, 58, 93, 99-100
Bed Shear Stress, 142-143, 149-150, 152-153, 157
Benthic Environmental Assessment Sediment Tool, 139
Blowouts, 23
Bottomset Beds, 33

C

Cementation, 53, 59, 78-80
Ceramic Material, 221, 224, 233
Chen 3d Software, 48
Clast Size, 108-109, 195
Clastic Rocks, 1, 16, 53-54, 73-74, 81, 135, 196
Coastal Flood Prevention, 47
Coastal Planning, 159
Cohesive Sediments, 157
Conglomerate, 51, 76, 105-108, 110, 196
Crystallinity, 171, 219-220

D

Debris Flow, 80-81, 107, 139
Deepwater Marine, 108
Defect Initiation, 18
Dimension Stone, 2, 234-236, 238-245, 258

E

Earth Surfaces Processes And Landforms, 27
Eocene, 63, 67, 189, 195, 201, 205-206
Eogenesis, 78-79
Epeiric Sea Phosphorites, 95
Erosion Function Apparatus, 48
Evaporites, 1-2, 58, 79, 132, 178, 188, 190, 247

E

Exner Equation, 6, 13, 27-28

F

Feldspar, 54-56, 62, 75, 77, 79-80, 84-86, 180
Flood Hazards, 41
Fluvial Processes, 7, 9, 94, 137
Foreset Beds, 33, 119
Fraser Delta, 34
Froth Flotation, 96

G

Gilbert Deltas, 31
Groynes, 47
Gully Erosion, 12, 162

H

Horizontal Soil Flux, 27

I

Indo-australian/sunda Collision, 201
Instantaneous Initiation, 18-19
Iron-rich Sedimentary Rocks, 53, 58, 97-99, 101
Ironstones, 58, 98-99, 104
Iso 14688-1:2002, 17

K

Krumbein Phi Scale, 16

L

Levee, 42-49
Levee Breach, 48
Lithospheric Compression, 186
Littoral Environments, 94
Littoral Sands, 13
Lower Flow Regime, 111-112, 124, 127

M

Mass Balance, 13, 148
Mesogenesis, 79
Milankovitch Cycles, 72
Miocene, 189, 193-198, 202
Mobile Bed, 18

Mud Transport, 86
Mud-dominated Delta, 87
Mudcracks, 66, 114
Mudrocks, 55-58, 74-75, 77, 79, 82-84, 86-91

N
Native Vegetation, 14
Null Point Theory, 14, 156, 158-159

O
Object Gouges, 114
Oligocene, 191, 201, 205-206
Overland Flow, 9, 11-12, 162

P
Particle Motion, 9, 140
Petroleum Geology, 2, 217-219, 246-248, 275
Phosphorite, 53, 91-92, 94-96
Phosphorus Cycle, 93, 95
Pliocene, 194-195, 197-199, 202
Point Bar, 35-38, 120

R
Ripple Marks, 20, 65-66, 100, 110-112, 123-124, 126-127
River Delta, 29, 32, 34, 47, 52

S
Sediment Accumulation, 13
Sediment Discharge, 139
Sediment Flux, 27-28
Sediment Size, 7, 156
Sedimentary Breccias, 80-81, 106
Selima Sand Sheet, 26

Sequence Stratigraphy, 4, 203, 209
Shelf Margin, 14
Shelf Slope, 14
Soft-sediment Deformation Structures, 113, 131-132
Sole Markings, 66, 114, 128-129, 132-133
Stony Deflation Zones, 23
Stress Balance, 140
Strike-slip Deformation, 186
Supratidal Environments, 94
Surface Runoff, 11, 161-162, 168
Suspended Load, 10, 21, 33, 38, 50, 58, 145-146, 148-149, 153-155, 157
Synthetic Aperture Radar, 41

T
Telogenesis, 80
The Bed Aggrades, 28
Tidal Flat, 72, 94
Tidal Flats, 21, 66, 68, 108, 159
Tide-dominated Deltas, 31
Topography, 3, 36, 105, 120, 168, 186, 188, 192, 196, 254

U
Upper Plane Bed, 20-21, 111

V
Vegetation Removal, 27

W
Watershed Development, 14
Wave-dominated Deltas, 31
Weathering Rocks, 87